AF595120

Advances in Microfluidics Technology for Diagnostics and Detection

Advances in Microfluidics Technology for Diagnostics and Detection

Editors

David Kinahan
Dario Mager
Elizaveta Vereshchagina
Celina Miyazaki

MDPI • Basel • Beijing • Wuhan • Barcelona • Belgrade • Manchester • Tokyo • Cluj • Tianjin

Editors
David Kinahan
School of Mechanical Engineering
Dublin City University
Dublin
Ireland

Dario Mager
Institute of Microstructure Technology
Karlsruhe Institute of Technology
Karlsruhe
Germany

Elizaveta Vereshchagina
Microsystems and Nanotechnology Department (MiNaLab)
SINTEF AS
Oslo
Norway

Celina Miyazaki
Centre of Science and Technology for Sustainability
Federal University of São Carlos
São Carlos
Brazil

Editorial Office
MDPI
St. Alban-Anlage 66
4052 Basel, Switzerland

This is a reprint of articles from the Special Issue published online in the open access journal *Processes* (ISSN 2227-9717) (available at: www.mdpi.com/journal/processes/special_issues/microfluidics_technology_diagnostics_detection).

For citation purposes, cite each article independently as indicated on the article page online and as indicated below:

LastName, A.A.; LastName, B.B.; LastName, C.C. Article Title. *Journal Name* **Year**, *Volume Number*, Page Range.

ISBN 978-3-0365-1366-9 (Hbk)
ISBN 978-3-0365-1365-2 (PDF)

Contents

Editorial

Special Issue on "Advances in Microfluidics Technology for Diagnostics and Detection"

David J. Kinahan [1,*], Dario Mager [2], Elizaveta Vereshchagina [3] and Celina M. Miyazaki [4]

1 School of Mechanical and Manufacturing Engineering, Dublin City University, Dublin D9, Ireland
2 Institute of Microstructure Technology, Karlsruhe Institute of Technology, Hermann-von-Helmholtz-Platz 1, 76344 Eggenstein-Leopoldshafen, Germany; dario.mager@kit.edu
3 Microsystems and Nanotechnology Department (MiNaLab), SINTEF Digital, SINTEF AS, Gaustadalléen 23C, 0373 Oslo, Norway; Elizaveta.Vereshchagina@sintef.no
4 Centre of Science and Technology for Sustainability, Federal University of São Carlos, Sorocaba, São Carlos, SP 13565-905, Brazil; celinammiyazaki@gmail.com
* Correspondence: david.kinahan@dcu.ie

Citation: Kinahan, D.J.; Mager, D.; Vereshchagina, E.; Miyazaki, C.M. Special Issue on "Advances in Microfluidics Technology for Diagnostics and Detection". *Processes* **2021**, *9*, 854. https://doi.org/10.3390/pr9050854

Received: 23 April 2021
Accepted: 6 May 2021
Published: 13 May 2021

In recent years microfluidics and lab-on-a-chip havecome to the forefront in diagnostics and detection. At the point-of-care, in the emergency room, and at the hospital bed or GP clinic, lab-on-a-chip offers the potential to rapidly detect time-critical and life-threatening diseases such as sepsis and bacterial meningitis. Furthermore, portable, cost-efficient, and user-friendly diagnostic platforms have a great deal of potential for global health applications. They can enable disease diagnostics and detection to occur in resource-poor settings where centralized laboratory facilities may not be available. At the point-of-use, microfluidics and lab-on-a-chip concepts can be applied in the field to rapidly identify plant pathogens, thus reducing the need for damaging broad-spectrum pesticides while also reducing food losses. Microfluidics can also be applied to the continuous monitoring of water quality and can support policymakers and protection agencies in protecting the environment. Perhaps most excitingly, microfluidics also offers the potential to enable the development of entirely new diagnostic tests that cannot be implemented using conventional laboratory tools. Examples of microfluidics at the frontier of new medical diagnostic tests include early detection of cancers through circulating tumor cells (CTCs) and highly sensitive genetic tests using droplet-based digital PCR.

The Special Issue on "Advances in Microfluidics Technology for Diagnostics and Detection" in *Processes* collects recent work from researchers active in this field. The Special Issue is available online at https://www.mdpi.com/journal/processes/special_issues/microfluidics_technology_diagnostics_detection. The interdisciplinary nature of research in the field of lab-on-a-chip has resulted in a collection of papers addressing a wide range of topical areas. For instance, Loy et al. [1] describe a low-cost and modular mixing platform for creating three-component siRNA polyplexes. They describe the use of low-cost microcontrollers (Raspberry Pi) to provide a flexible and programmable platform. The integration of low-cost microcontrollers with microfluidics is an emerging trend to enable distributed (and often wireless) sensing. The wireless nature of these concepts simplifies sealing and contamination issues, which justifies the extra effort. Many properties can today be measured wirelessly, as shown in Toto et al. [2] where they describe the development and miniaturization of a wireless vacuum sensor, which is enabled by combining the Pirani principle and surface acoustic waves (SAWs).

The COVID-19 pandemic has had a major impact on the economy and the overall health of the world. It has also had a great impact on the inter-connectivity of the globe. The rapid development of effective vaccines has demonstrated the capability of the scientific community to respond quickly and effectively to such challenges. Similarly, in the field of microfluidic diagnostic and detection, much work has been undertaken to improve the efficacy of COVID-19 diagnostic testing. While vaccine production is ramped up, diagnostic 'test and trace' will remain the primary tool healthcare officials have to attenuate the spread of the disease. Centralized laboratory testing and the development of low-cost portable tests for use in the field have been identified as the two major pillars for enabling 'test and trace'. With reliable field-based testing still some distance away, the characterization of laboratory-based assays is perhaps even more critical in the short-term. Towards this, Xie et al. [3] demonstrate detection of the SARS-CoV-2 virus, at low viral load estimates, using a commercially available microfluidic qPCR laboratory instrument.

In the field of point-of-care testing, the centrifugal lab-on-a-disc (LoaD) platform is identified as having great potential due to its portability, flexibility, and relatively low cost. This technology has been applied to a wide range of applications. Hin et al. [4] present a portable laboratory instrument that can be deployed in resource-poor settings to support epidemiological control of diseases (i.e., malaria, zika, etc.) associated with the mosquito vector. Importantly, the platform can also test for markers associated with an insect's species and its resistance to insecticides.

Centrifugal microfluidics also has potential in enabling and automating new diagnostic tests in the laboratory. In recent years, digital PCR has emerged as a highly sensitive diagnostic test for diseases of genetic origin. However, it can be difficult to implement in the laboratory and requires specialized and often expensive laboratory instrumentation. Schlenker et al. [5] use a centrifugal approach to demonstrate four-plex digital droplet PCR (ddPCR) capable of detecting 3.5–35 mutant DNA copies from 15,000 wild-type DNA copies for application in the detection of point-mutations associated with cancers. Finally, as a reflection of the emerging importance of the lab-on-a-disc platform, Miyazaki et al. [6] present an extensive review of biosensing technologies for use in centrifugal microfluidics.

Funding: This editorial received no external funding.

Conflicts of Interest: The authors declare no conflict of interest.

References

1. Loy, D.M.; Krzysztoń, R.; Lächelt, U.; Rädler, J.O.; Wagner, E. Controlling Nanoparticle Formulation: A Low-Budget Prototype for the Automation of a Microfluidic Platform. *Processes* **2021**, *9*, 129. [CrossRef]
2. Toto, S.; Jouda, M.; Korvink, J.G.; Sundarayyan, S.; Voigt, A.; Davoodi, H.; Brandner, J.J. Characterization of a Wireless Vacuum Sensor Prototype Based on the SAW-Pirani Principle. *Processes* **2020**, *8*, 1685. [CrossRef]
3. Xie, X.; Gjorgjieva, T.; Attieh, Z.; Dieng, M.M.; Arnoux, M.; Khair, M.; Moussa, Y.; Al Jallaf, F.; Rahiman, N.; Jackson, C.A. Microfluidic Nano-Scale qPCR Enables Ultra-Sensitive and Quantitative Detection of SARS-CoV-2. *Processes* **2020**, *8*, 1425. [CrossRef]
4. Hin, S.; Baumgartner, D.; Specht, M.; Lüddecke, J.; Mahmodi Arjmand, E.; Johannsen, B.; Schiedel, L.; Rombach, M.; Paust, N.; von Stetten, F. VectorDisk: A Microfluidic Platform Integrating Diagnostic Markers for Evidence-Based Mosquito Control. *Processes* **2020**, *8*, 1677. [CrossRef]
5. Schlenker, F.; Kipf, E.; Borst, N.; Paust, N.; Zengerle, R.; von Stetten, F.; Juelg, P.; Hutzenlaub, T. Centrifugal Microfluidic Integration of 4-Plex ddPCR Demonstrated by the Quantification of Cancer-Associated Point Mutations. *Processes* **2021**, *9*, 97. [CrossRef]
6. Miyazaki, C.M.; Carthy, E.; Kinahan, D.J. Biosensing on the Centrifugal Microfluidic Lab-on-a-Disc Platform. *Processes* **2020**, *8*, 1360. [CrossRef]

Review

Biosensing on the Centrifugal Microfluidic Lab-on-a-Disc Platform

Celina M. Miyazaki [1,*], Eadaoin Carthy [2] and David J. Kinahan [2]

[1] Centre of Science and Technology for Sustainability, Federal University of Sao Carlos, Sorocaba, Sao Paulo 18052-780, Brazil

[2] School of Mechanical and Manufacturing Engineering, Dublin City University, Glasnevin, Dublin 9, Ireland; eadaoin.carthy@dcu.ie (E.C.); david.kinahan@dcu.ie (D.J.K.)

* Correspondence: celinamiyazaki@ufscar.br or celinammiyazaki@gmail.com

Received: 17 September 2020; Accepted: 22 October 2020; Published: 28 October 2020

Abstract: Lab-on-a-Disc (LoaD) biosensors are increasingly a promising solution for many biosensing applications. In the search for a perfect match between point-of-care (PoC) microfluidic devices and biosensors, the LoaD platform has the potential to be reliable, sensitive, low-cost, and easy-to-use. The present global pandemic draws attention to the importance of rapid sample-to-answer PoC devices for minimising manual intervention and sample manipulation, thus increasing the safety of the health professional while minimising the chances of sample contamination. A biosensor is defined by its ability to measure an analyte by converting a biological binding event to tangible analytical data. With evolving manufacturing processes for both LoaDs and biosensors, it is becoming more feasible to embed biosensors within the platform and/or to pair the microfluidic cartridges with low-cost detection systems. This review considers the basics of the centrifugal microfluidics and describes recent developments in common biosensing methods and novel technologies for fluidic control and automation. Finally, an overview of current devices on the market is provided. This review will guide scientists who want to initiate research in LoaD PoC devices as well as providing valuable reference material to researchers active in the field.

Keywords: biosensors; LoaD platforms; microfluidics; centrifugal microfluidics; PoC devices

---

1. Introduction

The development of microfluidic biosensors with rapid, sensitive, and selective response has been the focus of many research laboratories. Biosensors can be focused both on the clinical diagnosis of silent diseases, such as cancer, to guarantee that the patient is referred to as quickly as possible to the appropriate treatment, and in the routine control of chronic diseases such as diabetes, to give a rapid and easy indication of the medication dosage. Biosensors are defined by the International Union of Pure and Applied Chemistry (IUPAC) as "a device that uses specific biochemical reactions mediated by isolated enzymes, immunosystems, tissues, organelles or whole cells to detect chemical compounds usually by electrical, thermal or optical signals" [1]. The biomolecule responsible for the analyte recognition is denoted as the biorecognition element [2], and is the key for the specificity of the biosensor. Since 2010, about 28,000 articles have been published in accordance with ISI of knowledge (Web of Science) with the topic "biosensors" (black bars on Figure 1A). Biosensors exhibit advantages over the traditional analytical methods including low-cost of manufacture, fast response time, easy handling (does not demand trained operators), portability, better batch-to-batch reproducibility, and often have comparable sensitivity and selectivity. Despite this, few biosensors have been commercialised due to the challenges with translating laboratory-based platforms to devices and instruments suitable for widespread application in practical situations.

Sample preparation is a critical step in ensuring that a biosensor can meet the required sensitivity and specificity. Sample preparation is often an arduous and labour-intensive task involving a range of laboratory unit operations (LUOs) such as metering, mixing, separation, resuspension, and elution. The versatility and usefulness of a biosensor is greatly reduced if sample preparation must be performed manually. Therefore, much effort has been made to encapsulate biosensors within microfluidic devices, or Lab-on-a-Chip devices, which can fully automate the required LUOs and therefore offer sample-to-answer performance. Microfluidics is related to the science and technology of systems that process or manipulates small amounts of fluids [3,4]. Biosensors based on microfluidic technology can offer reduced reagent volumes, reduced sample volume, increased automation, and can function without requiring expensive and large instrumentation. These advantages, combined with the potential for mass manufacturing, can lead to low-cost tests and can prevent sample and operator contamination by minimising the user intervention [5]. The combination of "biosensors" and "microfluidics" has been a consistent focus of the research community as shown in Figure 1A (connected blue squares).

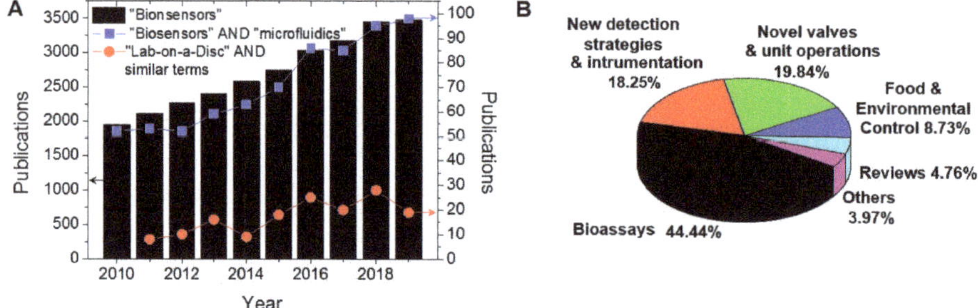

Figure 1. Web of Science results analysis by publication year using the topic (**A**) "biosensors" (black bars), the combination of "biosensors" and "microfluidics" (connected blue squares), and the terms "lab-on-a-disc" or "lab-on-a-CD" or "bio-disk" or "lab-CD" or "lab-disc" (connected red circles). (**B**) "lab-on-a-disc" results distributed in applications focus. Data from Clarivate Analytics (ISI Web of Knowledge [6]) in 18 May 2020.

Flow control (via pumping and valving) is a critical technology for enabling microfluidic systems. Prior to undertaking the design and fabrication of LoaD technology, it is important to understand non-centrifugally induced fluid flow mechanisms. Many methods of fluid manipulation and delivery have been reported. Passively driven microfluidics is a popular method as it is operated without an external fluid delivery system or actuator. Techniques such as osmosis [7,8], capillary action [9], pressure [10], gravity-driven flow [11], surface tension [12], vacuum-driven pressure [13], and hydrostatic flow [14] techniques can all be applied to achieve adequate fluid flow within microchannels. Applying these methods to microfluidic devices creates a robust, self-sufficient platform which can be utilised in any POC setting. Applying external triggers can further the capabilities of the devices and incorporates a variety of active driven device techniques. Manipulating fluid using an external trigger can allow for more complicated assays to be carried out in situ. Active fluid flow methods include electroosmotic [15], acoustic [16], photo-actuated [17], and pumping [18]. Both passive and active pumping mechanisms are important in understanding the dynamics of microfluidic devices. These methods may also be applied to centrifugal devices and can be amalgamated to create highly comprehensive POC tools. Centrifugal platforms can incorporate both types of fluid flow techniques; however, centrifugal forces play the main role of fluid displacement within the platform architecture. The Lab-on-a-Disc (LoaD) platform uses a typically disc-shaped cartridge and the pumping is via the centrifugal force; the disc-shaped chip is rotated about its axis. The general advantages of polymer chip-based microfluidics are that (i) mass production is possible with non-expensive infrastructure and common polymers; (ii) low-cost acrylics or thermoplastics have replaced glass for chip production, and, therefore, the cost per unit is significantly reduced compared to other in vitro diagnostic (IVD) tests [19]; aside from those advantages, the LoaD systems present particular advantages, specifically that (i) the fluidic propulsion depends only on a low-cost and controllable spinning motor, without the need of pneumatic interfaces and pumps; (ii) liquid handling is widely independent of the sample properties, i.e., pH or ionic strength (pivotal for electrokinectic-based methods); (iii) possible full integration, automation, and miniaturisation are possible; and (iv) there is possible large-scale parallelisation and multiplexing of bioanalytical assays. Because of the optical transparency of polymer LoaDs and the no-contact nature of the detection set-up, the optical detection has been the most common approach in LoaD biosensors. Expertise developed in manufacture in disc-based storage media (Compact Disc ™ and Digital Versatile Disc ™) provides a strong knowledge base to support process development for manufacture of LoaDs. Figure 1A (connected red circles) illustrates the number of publications concerning LoaD platforms. Because different terms have been used in the literature, the search was made using "Lab-on-a-disc" OR "lab-on-a-CD" OR "Bio-disk" OR "Lab-disc" OR "Lab-CD" as topic keywords. The number of publications are generally trending upwards demonstrating an increased interest in the LoaD platform while the variability in the trend likely reflects that LoaD remains a niche platform. The application focus distribution of LoaD publications is shown in Figure 1B.

Recent developments in microfluidic research has allowed for the integration of well-established detection methods, such as optical or electrochemical, to create robust platforms capable of a sample-to-answer within a significantly reduced time [20]. The use of centrifugal microfluidic platforms allows for complex and automated sample handling integrated within the microfluidic chip. Alongside the favourable features listed above, a key advantage of the LoaD platform is the capability to apply centrifugation to the preparation of the sample for the biosensor. This is particularly useful when processing complex sample matrices such as whole blood; here, for example, it can be separated into its constituent components (plasma, red blood cells, and white blood cells) in a single step [21]. Furthermore, the reduced costs of micro-processors has also resulted in the emerging area of electronic LoaDs (eLoaD) where the instrumentation required by the biosensor can be miniaturised into a support instrument and, in some cases, can co-rotate with the LoaD. This allows for detection and analysis of biomolecules with significantly reduced data acquisition times [22]. Contextualising, for the majority of electrochemical-LoaDs, the electrodes are integrated onto the platform to connect the biosensors to the instrumentation while the electrochemical readouts are often made from a stationary disc [23,24]. Recently, measurements during disc rotation were enabled by electrical slip-rings [25] or, where the high background electrical noise associated with some slip-rings may greatly interfere with the electrical readout signal, using integrated micro-controllers with wireless data transfer [26,27].

Lab-on-a-Chip platforms (which are often called micro total analysis systems (µTAS)) that are based on LoaD platform are particularly suitable for point-of-care (PoC) diagnostics in low-resource settings. Application of these platforms in poor and remote areas is termed extreme point-of-care [28]. Guidance from the World Health Organization (WHO) recommends that the development of diagnostic tests to resource-limited settings should follow the ASSURED (Affordable, Sensitive, Specific, User-friendly, Rapid/Robust, Equipment-free, Deliverable) criteria [29,30]. An updated acronym was proposed in 2019 as REASSURED, with the inclusions of R as Real-time connectivity and E as Ease of specimen collection and Environmental friendliness [31]. Researchers have made great efforts to meet all these criteria but challenges still remain [19,28,32]. Certainly, centrifugal microfluidics has features which position it as the ideal platform for PoC diagnostic testing and is, therefore, at the forefront of extreme PoC device development.

The existing academic literature contains a number of well-regarded review papers addressing both the general LoaD platform and specific aspects of this technology platform [5,28,33–39]. In this review, we discuss the recent advances in centrifugal microfluidics focusing on biosensing applications. We tried to concentrate our discussion (but do not limit) in the literature over the past 10 years. We divide this paper into sections concerning the main protocols in biosensors. We aim to provide an accessible guide for the ample audience of biosensor researchers with interest in leveraging LoaD technology to enhance their existing research while also providing a valuable resource for microfluidic specialists. Section 2 introduces the fluidic control in LoaD biosensors. Section 3 discusses the LoaD manufacture pointing both polymer microfabrication and biorecognition element immobilisation strategies. Section 4 describes the essential processing in LoaD biosensors while Section 5 discusses the existing and emerging LoaD PoC devices. We finalised with future perspectives and our final remarks in Sections 6 and 7, respectively.

2. Fluidic Control in LoaD Biosensors

The main driven-force in LoaD biosensors is the centrifugal force which acts outwards from the centre of rotation. Other pseudo-forces which play a role are the Coriolis force (applied when a particle/fluid moves in a rotating reference frame) and Euler force (induced by accelerating or decelerating the disc rotation). A combination of channel design, capillary action, and the placement of valves control can control the movement and timing of fluid movement. When a disc spins, the centrifugal force induces fluid flow radially outwards from the centre to the edge of the disc. Considering a fluid of mass density ρ on a planar disc rotating at a distance r from the central axis at an angular velocity ω, this liquid experiences a centrifugal force f_ω (Equation (1)) [34,35,40]

$$f_\omega = \rho r \omega^2 \tag{1}$$

a Euler force f_E given by Equation (2) [34,35]

$$f_E = \rho r \frac{d\omega}{dt} \tag{2}$$

with a rotational acceleration $\frac{d\omega}{dt}$, and a Coriolis force f_C (Equation (3)) [34,35,40]

$$f_C = 2\rho\omega v \tag{3}$$

where v is the fluid velocity in the plane. These three forces are represented in Figure 2, and can be controlled by the frequency of the rotation ω [35]. The centrifugal force induces the fluid to flow radially outward from the centre of the disc to the outer circumference [40]. The Coriolis force depends on the direction of rotation. If the direction changes, the direction of the Coriolis force changes [41]. It can be applied to mixing of samples, for flow switching, or directing of sample in specific channels [40]. In the "shake-mode" with continuous change in the spin speed, the angular momentum caused by the acceleration and deceleration induces Euler forces, resulting in a layer inversion of the fluid in the microfluidic chamber [34].

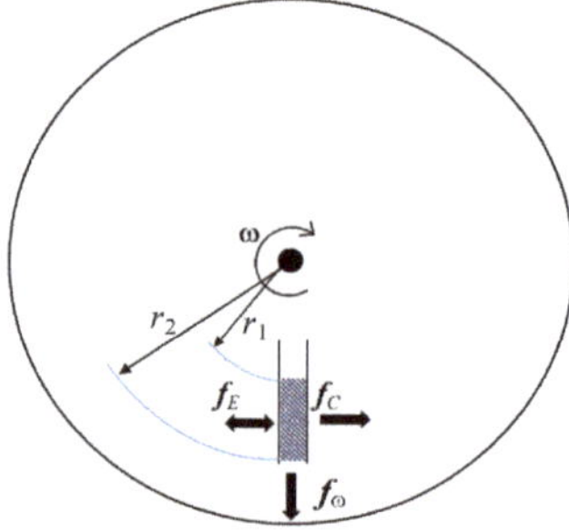

Figure 2. Forces acting in centrifugal microfluidics. The centrifugal force f_ω acts radially outward, the Coriolis force f_C acts perpendicular to ω and fluid speed, and the Euler force f_E is proportional to the angular acceleration.

The centrifugal flow rate depends on the rotational speed, radial location of the fluid reservoirs, channel geometry, and fluid properties such as viscosity, density, etc. [36]. Physicochemical properties as pH and ionic strength have no significant influence on the centrifugal flow rate, making it possible to pump many different fluids and integrate various processes on the same disc.

In a non-rotating reference frame under gravity (i.e., on the surface of the earth), the hydrostatic pressure is defined by the simple equation $\Delta P = \rho g h$ where g is the acceleration due to gravity and h is the height of the fluid column. Analogously, the centrifugally induced hydrostatic pressure is dependent on the radially inward and radially outward locations of a liquid columm on the disc (defined by r_1 and r_2 respectively in Figure 2). The centrifugally induced pressure, ΔP_c is defined as

$$\Delta P_c = \rho \omega^2 (\frac{r_2^2 - r_1^2}{2}) \tag{4}$$

In literature, this equation is commonly represented in an alternative form

$$\Delta P_c = \rho \omega^2 \Delta r \bar{r} \tag{5}$$

where Δr is the radial height of the liquid column ($\Delta r = r_2 - r_1$) and $\bar{r}$ is the radial center of the liquid column ($\bar{r} = \frac{(r_2 - r_1)}{2}$) [5].

While rotating, the centrifugal force acts on all liquids present on the disc and, therefore, to automate common assays, valves are required. Valves act just like faucets that control if and when the channels will deliver the fluid. Different types of valves have been proposed in the past years ranging from simple designs to highly sophisticated implementations. On the LoaD valves are typically "normally closed" (NC) such that in their initial state, they are closed and will open with some stimulus. Valves can also be "normally open" (NO), but these are less common and useful on LoaD. Most valves are classified as "single use" (i.e., changing from closed state to open state), but others can be changed between open and closed states. Valves are also classified as rationally-actuated or instrument-actuated [42]. Rotationally-actuated valves (also called passive valves) are controlled by changes in the disc spin rate (centrifugal force) while instrument-actuated valves (also called active valves) use an external actuator or source of energy (e.g., a laser, a pneumatic pump, a robotic arm) to control on-disc valves. The different types of valves and flow control techniques have recently been thoroughly reviewed [34,43]. However, here, we will introduce some of the most common and relevant valves to demonstrate how they are key components in the integration and parallelisation of bio-analytical processes on LoaD platform.

Capillary valves (see Figure 3A) are based on the balance between the capillary pressure and the centrifugally induced pressure [36]. In a common configuration, the capillary burst valve is connected to a reservoir through a straight channel, and when the disk is at rest or in low rotation speed, the liquid flows to fill the hydrophilic channel. However, it stops at the inlet of a suddenly expanded valve due to the capillary pressure [44], so the liquid will pass through a capillary valve only when the centrifugal force is high enough to break the capillary barrier pressure. In the hydrophobic valve (Figure 3B), hydrophobic regions in the microchannel prevent the fluidic movement. The valve opens when the centrifugal force overcomes a certain critical value [34,36]. In a siphon valve (Figure 3C), the disc is rotating in a high speed and the centrifugal force keeps prevents priming (via capillary pumping) of the siphon. With the decrease of the rotation speed, the siphon's hydrophilic channel is primed by the capillary force (acting against the direction of the centrifugal force). The liquid in the siphon is pumped by the capillary force over the siphon crest and radially outwards until the liquid meniscus passes the radial inward location of the liquid column (i.e., location r_1 described in Figure 2). At this point, centrifugal pumping far exceeds capillary pumping and the siphon valve opens when emptying of the reservoir can be sped up by increasing the disc spin rate/centrifugal pumping force [36]. Variations of siphon valves including centrifugo-pneumatic siphon valves [45], sequential siphon valves [46,47], and interruptible siphon valves [48] can also be used for additional control.

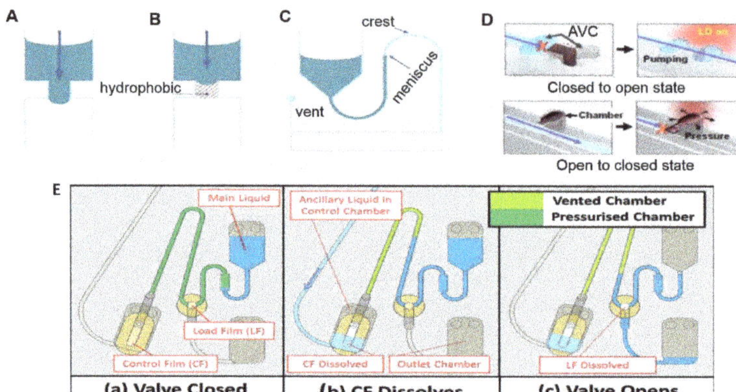

Figure 3. Centrifugal microfluidics valving methods: (**A**) capillary valve on a hydrophilic microchannel; (**B**) hydrophobic and (**C**) siphon valves. Adapted from [36] with permission from The Royal Society of Chemistry. (**D**) Laser irradiated ferrowax microvalves: at the top, from a closed to open state, the laser beam is focused at the ferrowax that melts and flows to an assistant valve chamber (AVC); at the bottom, to block an open channel, the laser is focused at the pre-loaded ferrowax located adjacent to the main channel. The molten ferrowax bursts into the main channel and solidifies, blocking it. Adapted from [49] with permission from The Royal Society of Chemistry; and (**E**) event-triggered valve based on dissolvable film (DF): (**a**) sample is loaded and the two DF tabs create a closed pneumatic chamber. The counteracting gas pressure keeps preventing the liquid to enter the pneumatic chamber; (**b**) the ancillary liquid enters and dissolves the control film (CF) and the pneumatic chamber is vented, allowing the main liquid to wet the load film (LF), and (**c**) sending liquid to the outlet chamber. Reproduced from [42] with permission from The Royal Society of Chemistry.

Sacrificial valves are another common valving technology. These valves involve the integration of some material which acts as a destructible barrier layer to allow liquid release. The release can be triggered externally, for example, by laser irradiation [49] (shown in Figure 3D). Laser irradiated ferrowax (iron oxide nanoparticles dispersed in paraffin) microvalves goes from a closed to open state by laser irradiating the ferrowax, which melts and flows to an assistant valve chamber (AVC). From open to closed state, the laser is focused at the pre-loaded ferrowax located adjacent to the main channel. The molten ferrowax bursts into the main channel and solidifies, blocking it [49]. Dissolvable films (DFs)-based valves [50–52] allow the passage of the fluid when its surface is wet, and can be used to implement an event-triggered system (Figure 3E) [42] which act analagous to a (single-use) electrical relay or transistor. DF valves are schematically positioned at specific sites on the disc, the arrival of liquid at this point triggers the release of liquid at another point [42]. In Figure 3E, the sample (main liquid) is loaded and the two DF tabs (named control film (CF) and load film (LF)) create a closed pneumatic chamber (dark green in (a)). The counteracting gas pressure keeps preventing the liquid to enter the pneumatic chamber, and thus wet the DF. At an appropriate timing, an ancillary liquid enters the control chamber and dissolves the CF, thus venting the pneumatic chamber (b), allowing the main liquid to wet the LF and finally, sending the liquid to the outlet chamber (c). This is particularly useful for sequential delivery steps when the arriving of a liquid in the waste chamber triggers the delivery of the next reagent/buffer. Many novel valving strategies have been described in the literature in the last years, always aiming to facilitate and automate disc protocols, as new passive [53–55] and active (magnetically induced [56], pinch-based valves [57], reversible [58–61], among others) valves. Some of them will be discussed in the next sections as they appear in the control of biosensors processes.

3. LoaD Fabrication

3.1. Polymer Microfabrication

At the end of the 1990s decade, polymers started to be used for microfabrication [62,63]. Opposite to silicon and glass, which are fragile and expensive, requiring time-consuming and complex processing, polymers are low-cost, of easy and scalable processing, and are available with a range of chemical and physical properties. LoaD fabrication techniques have followed the same strategies than conventional polymer-based chips. The choice of the material accounts on the cost, but especially on the desired properties, e.g., on-disc colorimetric measurement demands optically transparent polymers, some types of valves demands flexible structures, intricate design demands polymers which can be processed by high resolution techniques. As many specialised reviews have dedicated on microfabrication techniques [64–66], this review does not intend to go deep in such subject but displays the main LoaD fabrication approaches and offer the address to find out more about it.

A LoaD biosensor shelters various LUOs and commonly a net of 3D channels and reservoirs are needed, which is obtained by stacking of independently processed layers bonded to each other. The LoaD architecture is created in a computer-aided design (CAD) software, and each layer is previously processed by techniques as replica moulding (soft-lithography), xurography, laser ablation, etc. Figure 4A illustrate (a) 1.5 mm thick-poly(methyl methacrylate) (PMMA) plates machined using a CO_2 laser ablation, and (b) 86 µm thick-pressure sensitive adhesive (PSA) processed by a knife-cutter. (c) After removing the protective covers, the LoaD is assembled by stacking the PMMA and PSA layers using an alignment jig and were rolled using a laminating machine. The side view is shown in (d). Although the cutting process is automated and scalable, the assembly stage is more difficult to be automated while the manual assembly of layers is time-consuming and laborious. Yet, researchers widely use this approach at the proof-of-concept stage because the prototyping is fast and without the need for mould fabrication.

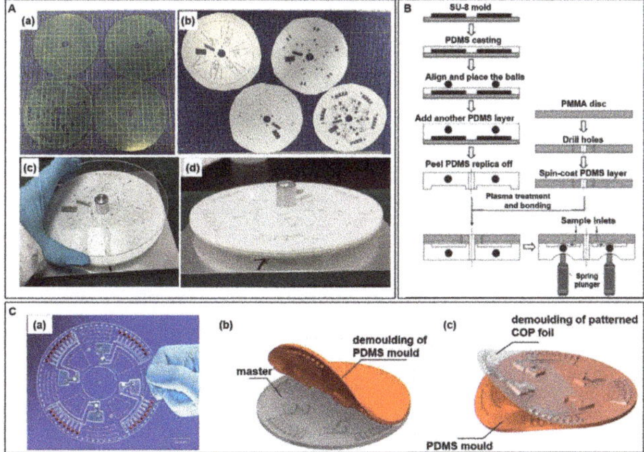

Figure 4. (**A**) Photographs of steps of layers assembly during the LoaD manufacture: (**a**) laser ablation cut PMMA discs, (**b**) knife-cut PSA layers, (**c**) stacking and alignment of layers, and (**d**) side view during the mounting of layers. (**B**) Schematics of PDMS moulding to fabricate the disc: Pinch valves-based disc fabricated by PDMS moulding. Reproduced from [57] with permission from Elsevier. (**C**) Hot embossing fabricated LoaD biosensor: (**a**) photograph of the LoaD, (**b**) fabrication of the PDMS mold using a milled PMMA master, and (**c**) demoulding of the hot embossed COP foil. Reproduced from [67] with permission from The Royal Society of Chemistry.

Polydimethylsiloxane (PDMS) is usually patterned by replica moulding (soft-lithography) in which the PDMS reagents (base + curing agent) is poured over the master mould and cured to solidify.

Commonly, the master mould is fabricated using UV photolithography of SU-8 (negative photoresist) exposing the pattern through a mask. The unexposed parts are dissolved while the cured photoresist remains on the substrate defining the pattern. Figure 4B illustrates a schematics of fabrication of a microfluidic disc by PDMS casting over the SU-8 master mould. At the design suggested by the authors to create pinch-valves, metal balls were placed over the first cured PDMS layer, and covered with a new layer of PDMS. The PDMS replica containing the metal balls are peeled off and united to a PDMS-coated and plasma-treated PMMA disc [57]. The active valves are controlled by spring plungers which push the balls deforming the flexible PDMS to block the flow [57]. Alternative low-cost techniques for master fabrication, as micromilling, laser engraving, cutting plotters, are discussed in [66]. PDMS is optically transparent in the visible range, biocompatible, flexible, and can be easily sealed to glass or another PDMS part (no need of adhesive layers). Soft-lithography provides high-resolution replicas (nm-scale), but pattern deformations and defects in the moulded product can occur in the peeling process [64]. Other disadvantages concern the cost, relatively high when compared to common thermoplastics [65], and the difficult to implement mass production [66]. Details in PDMS micromolding can be found in [68,69].

Hot embossing and injection moulding of thermoplastic materials are remarkable concerning cost-efficient mass production [70–72]. At hot embossing processing, the temperature is elevated above the polymer glass transition temperature (T_g), and the mould is pressed against the polymer substrate transferring the pattern [73]. Nanometre range can be achieved by hot embossing [71]. As polymer pattern resolution depends on the mould resolution, delicate patterns usually demand lithography/etching processed silicon moulds [70,74] accompanied by time-consuming and expensive fabrication [64]. In addition, the masters have limited cycles of life due to cracking and warping from the stress and temperature conditions [75]. To overcome these drawbacks, μ-scale PDMS patterns were produced by photolithography to act as hot embossing mould [75,76]. PDMS moulds are relatively inexpensive, rapid to fabricate and reusable for many replications, and possess thermal stability that enables to stamp common thermoplastic polymers [75,76]. Figure 4C present (a) a photograph of a cyclic olefin polymer (COP) foil disk, fabricated by hot embossing and sealed with PSA, and representations of (b) demoulding of the PDMS mould from the milled PMMA master, and (c) demoulding of patterned COP foil after hot embossing, cooling, venting, and opening of the chamber [67].

3D printing is an emerging technology in microfluidics [77]. Yet, it has some limitations concerning resolution. Usually, supporting material is needed to fill the void spaces during the printing process. Later, its removal is manually made which is time-consuming and difficult to do in small channels. The solvent assisted removing is mostly hampered because the chemical composition of supporting and construction materials are often similar [78]. The produced rough surfaces may also cause dead volumes and irregular surface modifications [78]. Concerning LoaD applications, 3D printing has been used to fabricated specific parts, as valves [61] or valve actuating discs [57,59], or in the manufacture of peripheral structures [25].

3.2. Immobilisation of the Biorecognition Element

The immobilisation of the biorecognition element is always a concern in the construction of biosensors because the sensitivity depends on the total activity of proteins. LoaD platforms, like most microfluidic devices, are usually built in polymer materials, and the direct passive adsorption of proteins as antibodies and enzymes onto these surfaces is mainly driven by hydrophobic interactions. The adsorption-induced conformation changes can lead to protein denaturation and reduce protein activity by as much as 90% [79,80]. Since hydrophobic surfaces induces higher denaturation than hydrophilic ones [81], some surface modification strategies have been proposed to increase the hydrophilicity of conventional polymers for microfabrication. These strategies include plasma treatment [82], coverage with layers of hydrophilic polymers [83,84] and graphene oxide [85]. Poly(ethyleneimine)—PEI introduces hydrophilic amine groups and acts as a spacer (increases the space between the biomolecules and the hydrophobic surface) and has been applied in PMMA [80,84,85] and cycloolefin [83] surfaces. PEI coating [80], PEI coating following oxygen plasma treatment [84], and nanostructured layer-by-layer film of PEI and graphene oxide [85] have all demonstrated capacity to improve ELISA performance on PMMA surfaces. Functionalisation of metal surfaces is also crucial for electrochemical and reflectivity-based approaches. Conventional methods based on thiol functionalisation using cysteamine and alkanethiols followed by NHS-EDC chemistry have also been applied to the LoaD [86,87].

The anchoring of bio-active beads within the LoaD to act for the biorecognition element anchoring is an alternative to direct immobilisation of biomarkers onto the flat polymer surface. Beads can be easily functionalised with the biorecognition element *a priori* off-chip and then be loaded to the microfluidic platform. This is particularly useful when the protein coating may cause difficulty sealing microfluidic chips when surfaces have been functionalised [88]. One of the most important advantages using beads is the higher surface to volume ratio (1 g of microbeads with a 0.1 μm-diameter has a surface area of 60 m^2) [89]. This can be achieved, for example, by a 3D column filled by the functionalised beads. Additionally, analytes and bioreagents can be easily transported in the fluidic system when attached to beads, by the centrifugal flow or by external forces as pressure-driven flow or especially, by the magnetic force [89,90]. Additionally, some protein-functionalised beads are commercially available or have well-established protein anchoring protocols. Immunoassays applications usually encompass the capture antibody anchoring via carbodiimide chemistry on carboxylated-beads [91,92], or by biotinylated antibodies on streptavidin-coated particles [93].

4. Essential Processes in LoaD Biosensing

4.1. Reagents and Sample Storage and Supply

At the R & D stage, the input of reagents in a LoaD platform is usually by manual loading, via pipetting, of the reagents and samples. However, long-term storage strategies, which minimises operator handling, reduces cold-chain requirements, reduces the need for specialised user-training, and which avoids contamination of both user and sample, are of critical importance. A number of different technologies have been demonstrated including stick-packages [94,95], glass ampoules [96], and elastic-membrane micro-dispensers [97].

Pre-storage of reagents in glass ampoules placed on LoaD platforms for DNA extraction was demonstrated with no loss of ethanol and water for 300 days at room temperature. Frozen storage was also possible without ampoule rupture. The release of liquids (buffers and ethanol) is made by a mechanical force (fingertip pressure) through the elastic lid of the cartridge. While the liquid is centrifugally displaced, a filter prevents the glass shivers to go forward [96]. van Oordt et al. [94] have developed stick-packs of aluminium/polyethylene composite foil aiming at long-term storage of reagents. The transverse frangible seal was fabricated by ultrasonic welding and it is adjustable for a specific burst pressure. The hermetically sealed stick-packs can store both liquid or dry reagents, with the first frangible seal (that separates solid and liquid) bursting at lower rotation frequency. After mixing, the second frangible seal bursts to release the mixture [94]. These stick-packs were further applied in a LoaD for nucleic acid-based detection of respiratory pathogens [95].

A long-term storage micro-dispenser was created by Kazemzadeh et al. [97] and tested in lab-on-a-chip and LoaD platforms. The micro-dispenser comprises a tube with a hole and an elastic membrane covering this hole. When the internal pressure is increased (by the centrifugal force) in enough amount to stretch the membrane, a path is provided for the liquid release. Figure 5A(a) illustrates the micro-dispenser placed on a disc. No liquid release happens while the centrifugal force does not exceed the membrane force. Increasing the internal pressure by the rotation speed, membrane stretches and releases the fluid (b). To implement this technology for blood separation, a micro-dispenser was created with a tube with two apertures, each one covered with different elastic materials (see Figure 5B(a)). The top and the bottom membranes are for plasma and blood cells delivery, respectively. The location of each aperture is well-planned, e.g., the first top aperture is located at a point above the level where the plasma is separated from blood cells. Figure 5B(b) illustrates three identical assays for blood separation on disc and (c) a single assay with a micro-dispenser placed on the disc with a piece of tape and a tacky adhesive. Because of the different elastic properties, each of the membranes releases the liquid at different rotation speeds. In this case, a higher centrifugal force is needed to release the blood cells (bottom membrane). The long-term stability of the micro-dispensers were demonstrated to be two and one year(s), for DI-water and ethanol 70%, respectively [97].

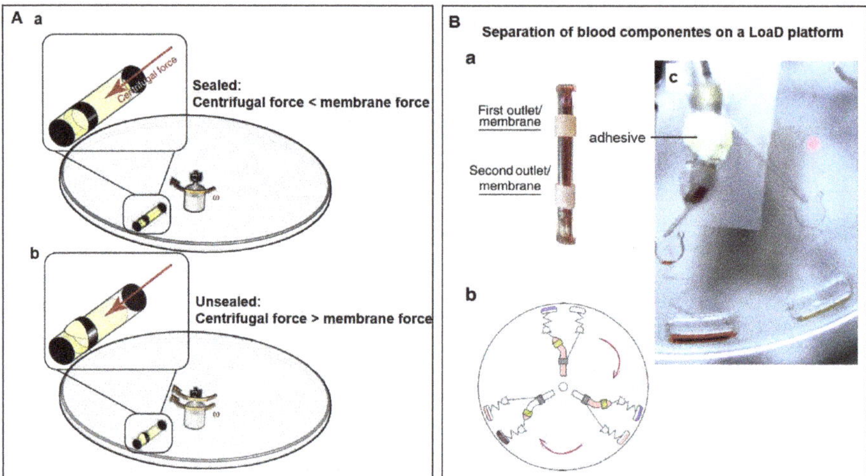

Figure 5. (**A**) Micro-dispenser working principle in a LoaD platform. (**a**) When the centrifugal force is lower than the membrane resistance, no liquid is released. (**b**) Increasing the rotation speed, the centrifugal force overcomes the membrane force and the fluid is temporary released. (**B**) Separation of blood on a LoaD platform. (**a**) Micro-dispenser with two apertures covered with two different membranes, C-flex and latex, and (**b**) a schematic of the LoaD platform for 3 identical assays for blood separation. (**c**) Picture of the disc dispensing blood plasma and blood cells to two different chambers at different rotation frequencies due to the difference in the membranes properties. Reproduced from [97] under a Creative Commons Attribution 4.0 International License.

4.2. Samples and Reagents Processing

4.2.1. Blood Processing

Blood is the most utilised sample for diagnostics, and its separation is usually the first step of many biological assays. Cell isolation and pathogens detection demand an initial step of selective blood separation. Conventional benchtop processes typically use centrifugation. This is relatively simple but often requires large sample volumes and, in case of infectious disease, there are concerns about operator safety. Based on the different densities of blood components, it is quite easy to promote separations simply by the rotation of the disc. Common protocols for plasma extraction on disc use moderate (about 1200 RPM for 3–8 min [98,99]) to fast rotation (2000–5000 RPM for few seconds [100–103]). Due to the difference in density of the plasma and the red and white blood cells, blood cells are driven toward the bottom while the plasma forms the supernatant layer on the top [101]. One of the great advantages of LoaD platforms is that the plasma extraction step can be easily and conveniently integrated with the following metering, mixing, and detection steps. It has been reported for plasma extraction, metering and mixing with reagents for the prothrombin (PT) time tests [101–103], nitrate/nitrite assays [98], cancer cells detection [86,104], virus detection [90] and many other sensing platforms.

Some geometry parameters can be defined to increase the speed of blood separation. Kim et al. [105] found that higher tilt angles and narrower channels promote faster plasma separation. The enhanced sedimentation in inclined channels is known as Boycott effect [106], and it is associated with the increase in the surface area available for the particles settling [105]. The impact of the Coriolis force on blood sedimentation has also been investigated [107]. *Spira mirabilis*-like structures (equiangular spiral) demonstrated an enhanced speed of sedimentation due to the increased Boycott effect, herein applied with a density gradient medium (DGM) [108]. Sedimentation over DGM uses solutions with specific densities, such as Ficoll and Percoll, to separate and sort blood cells into layers according to their inherent density differences [109]. This can facilitate the collection of the separated components, and thus the analysis of multiple blood samples. Figure 6A(a) illustrates the density distribution of human blood cells and (b) a schematic of the separation using Ficoll (density = 1.077 g mL^{-1}), with the dense cells moving to the bottom while the low density-cells remain on the top layer. Efforts have been done to develop DGM-based blood fractionation and separation of specific cells [21,86,104,110–113]. Morijiri et al. [113] proposed an elutriation-based separation in a centrifugal disc. Particles are introduced in the separation chamber with a low-density fluid under centrifugation. While particles experienced a centrifugal force in an outward direction, the fluid force acts in an inward direction. The retention position of each particle depends on the size and density, and are controlled by the balance of the centrifugal force and the fluid drag force. Small and low-density particles are not retained in the separation chamber and are the first to move towards the outlet. By introducing solutions with higher densities, the balance of forces is changed allowing a step-wise elution and recovery of particles [113].

Leukocytes extraction can be particularly important. Abnormal leukocytes counts can be related to anemia, infection, inflammatory conditions, and certain cancers [114]. Furthermore, they can interfere with the detection of circulating tumor cells (CTCs) [104]. CTCs have large size variation and their size can overlap with leukocytes causing unreliable size-based separations. Because CTCs are disseminated from primary tumors or metastatic sites to the bloodstream, LoaD platforms to separate them from whole blood have been developed [115,116]. Park et al. [104] developed a centrifugal platform with the selectivity based on the anti-EpCAM covered microbeads, which specifically bind to the CTCs, and thus make them heavier than other blood cells. This leads to sedimentation through a DGM and thus being isolated. Figure 6B shows the disc designed for sorting CTCs (left). The blood sample and microbeads conjugated with anti-EpCAM were injected via the blood chamber inlet while the DGM (Percoll) was pipetted through the DGM chamber inlet. Blood and microbeads are shown in the blood chamber (a), and after plasma separation, the triangle obstacle structure (TOS) played a role to retard the convection of the separated blood cells in the blood cell zone while the disc is stopped to open valve 1—V1 (to remove plasma to the waste chamber) (b). After the complete removal of plasma (c), V2 is closed to initiate the bead-binding (incubation) process by mixing (d). V3 is opened and the sample was transferred to the DGM chamber, where under centrifugation the microbeads-CTCs complexes were moved to the collection chamber, passing through the DGM layer due to their higher density, while the other cells remain on the top (e). After the complete separation, V4 is closed before the collection to avoid contamination (f) [104]. Here, the valving was based on the laser irradiation ferrowax [49], discussed previously.

Filtration-based strategies have also been successfully applied for blood separation. Sequential and tangential flow filtration using two track-etched polycarbonate membranes were integrated in the chambers by Kim et al. [117]. The pore size of the filters was carefully selected considering the size of platelets and blood cells. Figure 6C(a) depicts a photograph of the platelet isolation disc highlighting the filters. Schematics in Figure 6C(b) shows Filter-I (3 µm) to eliminate both white blood cells (8–15 µm) and red blood cells (6–8 µm). The plasma runs through Filter-II (600 nm) which captures platelets while washing residual plasma contents out to the waste chamber, as extracellular vesicles, proteins, lipids, and cell-free DNA [117]. Platelet analysis revealed high purity (>99%), free of white blood cells contamination, and the yield of the platelets recovery was significantly superior (>fourfold) than that conventional benchtop centrifugation [117]. Refer to the source for more details [117]. Table 1 summarises the blood separation and analysis integrated into LoaD platforms.

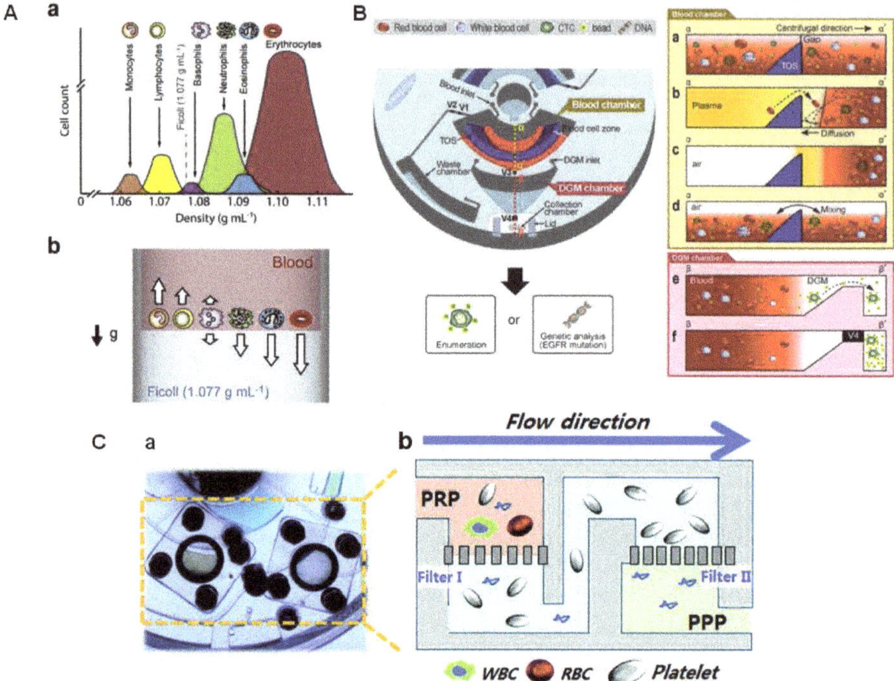

Figure 6. (**A**) Sorting of blood cells based on the DGM centrifugation. (**a**) The density distribution of human blood cells and (**b**) the movement of blood cells upon centrifugation with blood layer placed on the top of the Ficoll solution. The arrows indicate the movement direction of cells. Reproduced from [109] with permission from Elsevier. (**B**) Disc design for sorting CTCs from whole blood. TOS: triangle obstacle structure; V1, V2, V3, and V4 are on/off operation valves (wax that changes from liquid to solid state) controlled by laser irradiation. (**a**) Cross-sectional view of the blood chamber after injection of blood and microbeads; (**b**) after plasma separation; (**c**) after removal of the plasma with opening V1; (**d**) during incubation to bind CTCs to the antibody-functionalised beads; (**e**) the cross-sectional view of the DGM chamber showing that only CTC-microbead complexes (with higher density than DGM) are moved to the collection chamber; (**f**) V4 is closed to prevent sample contamination during collection. Reprinted with permission from [104]. Copyright 2014 American Chemical Society. (**C**) Filtration-based platelet isolation disc: (**a**) photograph and (**b**) schematics. Filter-I to eliminate both white blood cells and red blood cells from a platelet-rich-plasma (PRP). PRP runs through Filter-II which captures platelets while washing platelet-poor plasma (PPP) out to the waste chamber. Reproduced from [117] with permission from The Royal Society of Chemistry.

Table 1. Summary of the blood separation strategies performed in LoaD platforms. WB: whole blood, DGM: density gradient medium, PT: prothrombin time, HBV: hepatitis B virus, PBMC: peripheral blood mononuclear cell, CTC: circulating tumour cells, CRP: C-reactive protein, IL-6: interleukin-6.

Ref	Portion of Interest	WB Volume (Extracted Volume)	Separation Protocol	Separation Approach	Achievement
[98]	Plasma	70 µL (10 µL)	3 min @ 1200 RPM	Simple centrifugation	Successfully applied to nitrate and nitrite detection
[100]	Plasma	1 µL	1.5–6.0 s @ 1800 RPM	Simple centrifugation	Plasma separation efficiency of 96 %
[101]	Plasma	2.0 µL (0.5 µL)	2.5 min @ 2200 RPM	Simple centrifugation	Successfully applied to PT test.
[102]	Plasma	14 µL (3 µL)	80 s @ 4000 RPM	Simple centrifugation	Successfully applied to PT test
[103]	Plasma	7 µL (2 µL)	90 s @ 5000 RPM	Simple centrifugation	Successfully applied to PT test
[86]	Cells	5 assays × 20 µL	5 min @ 1500 RPM	DGM based separation and antibody functionalised sensor	Cell capture efficiency of 87% and CTC detection via EIS
[104]	Cells	5 mL	640 s @ 3000 RPM	Geometry + DGM	90% recovery of CTC isolation
[90]	Serum	500 µL (200 µL)	90 s @ 60 Hz	Simple centrifugation	Recovery of 90%. Successfully applied to PCR HBV detection
[21]	PBMC	18 µL	60 Hz	DGM	34% efficiency compared to hospital laboratory data
[110]	PBMC	8.75 µL	4 min @ 45 Hz	DGM	73% PBMC extraction (27 % below hospital laboratory data)
[111]	Cells	3 µL total (WB + beads)	40 min @ 10 Hz	centrifugo- magnetophoresis	92% efficiency of cells isolation
[112]	Blood fractions	10 µL	4 min @ 2700 RPM	Simple centrifugation with capillary valves to keep layers separated	95.15% leukocytes extraction
[113]	Blood fractions	20 µL diluted WB	720 s (+ 900 s) @ 2000 RPM (for erytrocytes elution/separation)	centrifugo- elutriation (balance between centrifugal force and fluid drag force)	98% recovery of erytrocytes
[117]	Plasma (platelet purification)	600 µL (150 µL platelet sample)	5 min @ 3000 RPM	Filtration-based isolation	99% purity with total assay time including washing = 20 min
[118]	Protein isolation	1 µL	5 min @ 8000–9000 RPM	DGM and antibody functionalised beads	CRP and IL-6 quantification (total assay time of 15 min)

4.2.2. Volume Metering and Aliquoting

Accurate metering of reagents and samples is an essential step in sample preparation to guarantee a reproducible and quantitatively reliable response from the biosensor. Although different architectures are found in the literature, the volume metering in LoaD is based on the chamber volume (controlled by its geometric parameters). Typically, an overflow channel is connected to this metering chamber to guarantee that any excess of liquid is sent to a waste/overflow chamber, as illustrated in Figure 7A. The same overflow principle can be applied to implement aliquoting and serial dilutions on disc, which are usually time-consuming and susceptible to errors (specially when working in µL-scale) when performed manually. As depicted in Figure 7B, Mark et al. [119] developed a metering structure whereby a dead-end pneumatic chamber enhanced system performance. In this structure, a single metering chamber is connected upstream and downstream by a feeding/overflow channel. The volume metering is based on centrifugo-pneumatic valve structure in which the metering chamber V_M is connected by a narrow channel to an unvented receiving chamber V_0 (a). The volume metering chamber is filled but the receiving chamber, which may, for example, contain lyophilised reagents, remains dry (b). When liquid reaches the narrow connection at the bottom, a meniscus is formed which prevents the air from escaping the unvented receiving chamber (creates a pressurised chamber with pressure p_1) (c). Aliquoting is performed by continuing the liquid feed. After complete the first V_M (d), the overflow channel deliveries the liquid for the next metering channel and so on, as illustrated in Figure 7C(a). After all the metering channels are completed, an increase in the centrifugal frequency reaches a point (burst frequency) in which the centrifugal force overcomes p_1, releasing the liquid to the receiving chambers, as shown in Figure 7B(e),C(b).

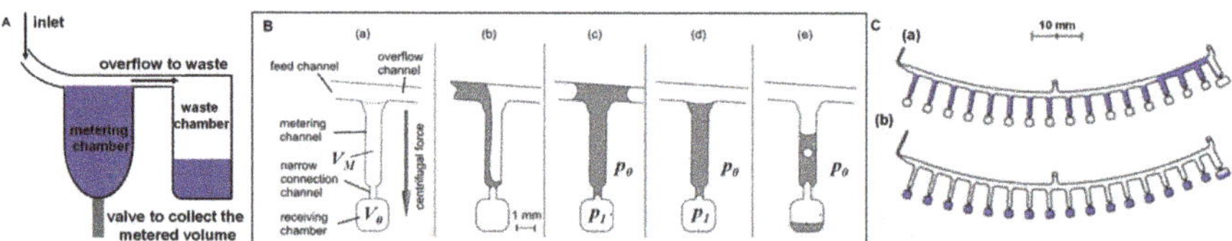

Figure 7. (**A**) Simple metering design based on overflow. (**B**) Metering structure based on the centrifugo-pneumatic valve: (**a**) design of a single metering structure with the volume defined by the metering channel V_M. (**b**) At a first centrifugal frequency, liquid fills the metering channel while air is displaced. (**c**) The pressure p_1 prevents the liquid to enter the receiving chamber V_0. (**d**) After metering, (**e**) an increase in the centrifugal frequency above the burst frequency (to an amount that overcomes p_1) releases the liquid. Adapted from [120] with permission from The Royal Society of Chemistry. (**C**) Layout of an aliquoting structure: (**a**) multiple V_M chambers been fed by liquid, and (**b**) aliquots distributed to the receiving chambers. Adapted by permission from Springer Nature [119], Copyright 2011.

Active valving based on individually addressable diaphragm (ID) valves [61] was used for on-disc serial dilutions by Kim et al. [121] (Figure 8A). These valves are assembled from an elastic epoxy diaphragm and a 3D printed actuator (Figure 8B). A simple push-and-twist action is required for closing and opening the channel. This open/close functionality allows the volume of sample and buffer to be metered in the two different zones; thus dictating the degree of dilution. These two zones are detailed in Figure 8C,D. An example of fivefold dilution is presented in Figure 8E (opened and closed valves represented by green and red colours, respectively): (i) at buffer metering zone, valve 1 (V1) is opened while V2, V3, V4, and V5 are closed for metering of total 9 µL-buffer. Simultaneously, at the sample metering zone, V6 and V7 are opened while V8 is closed, totaling 8 µL sample metered; (ii) at the sample metering zone, V6 is closed and the 8 µL-metered sample is sent to R1 (with valves 7, 8, 9, and 10 opened while V11 is closed). From the initial 10 µL sample, 2 µL remains in the mixing chamber. At the buffer metering zone, only the V5 is kept open, sending the buffer excess to the waste chamber; (iii) the 2 µL sample in the mixing chamber is diluted by the delivery of 8 µL buffer (only V2 and V3 are opened while V4 is kept closed keeping 1 µL buffer retained). Finally, the 5-fold diluted solution is metered (5 µL, by keeping V6 opened while V7 is closed) and sent to R2. The next dilution can be performed to the remained solution in the mixing chamber based on the similar steps (protocols for 2-,5-, and 10-fold dilutions are summarised in the table of Figure 8E). Buffer can be re-loaded by opening V1 [121]. Although the system at first appears complex and unwieldy, the recent miniaturisation and cost-reduction in micro-controllers means this approach can be easily automated and can flexibly address a wide array of biosensor applications.

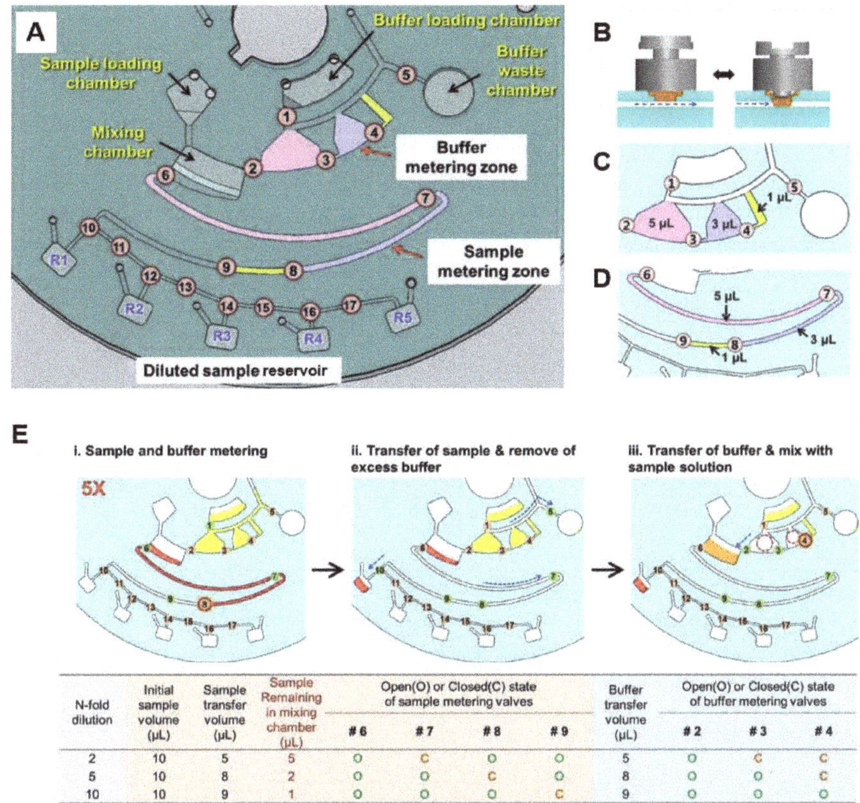

N-fold dilution	Initial sample volume (µL)	Sample transfer volume (µL)	Sample Remaining in mixing chamber (µL)	Open(O) or Closed(C) state of sample metering valves				Buffer transfer volume (µL)	Open(O) or Closed(C) state of buffer metering valves		
				#6	#7	#8	#9		#2	#3	#4
2	10	5	5	O	C	O	O	5	O	C	C
5	10	8	2	O	O	C	O	8	O	O	C
10	10	9	1	O	O	O	C	9	O	O	O

Figure 8. (**A**) Disc design for serial dilution consisting of sample and buffer loading chambers, mixing chamber and buffer waste chamber, two metering zones (for sample and buffer), five collecting reservoirs (R1–R5) for the diluted samples and seventeen ID valves (numbered circles). (**B**) A reversible ID valve composed of a diaphragm (orange) embedded on a top layer and 3D printed valve actuator (gray). Simple push and twist actions open and close the channel. Detailed buffer (**C**) and sample (**D**) metering zones. (**E**) 5-fold dilution steps: (i) in the buffer metering zone, a total of 9 µL is metered keeping V1 opened while V2, V3, V4, and V5 are closed. At the sample metering zone, V6 and V7 are opened while V8 is closed, totaling 8 µL metered; (ii) at the sample metering zone, V6 is closed and the metered liquid within the channel is sent to R1 (V7, V8, V9, and V10 are open while V11 is closed). At the buffer metering zone, V1, V2, V3, and V4 are closed and V5 is opened to send the excess to the waste chamber; (iii) 8 µL buffer is delivered (just V2 and V3 are open while V4 is kept closed keeping 1 µL of buffer retained) into the mixing chamber, where remains 2 µL sample. Thus, a dilution of 2 µL sample to a final 10 µL volume solution was achieved. The table presents the summary of the metered volume of sample and buffer, and the state of the valves to achieve 2-, 5-, and 10-fold dilutions. Reproduced from [121] with permission from Elsevier.

4.3. Mixing and Washing

Mixing protocols should provide an effective and fast homogenisation of the fluids. However, adaptation of simple bench-top protocols as agitation, vortexing and stirring onto the LoaD can cause unwanted opening of rotationally actuated valves; and the subsequent uncontrolled pumping of liquid. Therefore, systems which require rigorous mixing must be carefully designed. Several studies have been focused on developing disc protocols to allow an integrated and adequate mixing. On-disc mixing can be passive, just based on the rotation control, based on intrinsic forces as Euler or Coriolis [122–124] or by means of control of geometry [125,126]; or can be active based on external perturbations, e.g., compression by external actuators [127] or use of magnetic-beads inside the disc in combination with permanent magnets [128,129]. Several micromixing technologies are discussed in [130].

One of the most utilised mixing protocol is the "shake-mode" mixing [124] comprising periodic changes in the sense of rotation and rapid changes of the spinning frequency. The difference in the angular momentum leads to the appearance of a shear force that drives an advective current within the liquid [124]. The mixing time was down to 3.0 s for 25 μL-volume while a mere diffusion-based mixing took about 7 min. Combining the "shake-mode" mixing with magnetic beads (pre-filled in the mixing chamber) periodically deflected by a set of permanent magnets resting in the lab-frame, an effective homogenisation was attained in only 0.5 s [124].

The use of pneumatic pressure to promote reciprocating flow-based mixing [131] was introduced later, and has been successfully applied in LoaD biosensors [87,132]. This scheme utilises an auxiliary pneumatic chamber, and the sequential decrease and increase in the centrifugal force causes the compression and decompression of air pushing the liquid through different chambers and thus allowing an accelerate mixing [131]. Figure 9A(a) depicts an immunoassay disc. The biorecognition element (antigen) is immobilised on the reaction chamber to specifically capture the primary antibody (sample) while the secondary antibody is to allow the colorimetric detection. The schematics is illustrated in Figure 9B(b). The sequence of images in (c) describes: (i) the sample stored in the loading chamber; (ii) acceleration causing air compression in the pressure chamber; (iii) by deceleration, the air expansion in the pressurised chamber pushes the liquid back to the upper chamber; (ii and iii) the increase and decrease in rotation speed (3200–7000 RPM) are reciprocated to effective and rapid mixing; (iv) the rotation frequency is increased to the maximum (7500 RPM); followed by (v) a rapid decrease to very low value (480 RPM). This increases the liquid level above the crest point, primes the siphon valve and sends liquid to the waste chamber, (vi) leaving an empty reaction chamber [132]. Pneumatic valving and mixing are reviewed in [133]. Another reciprocating mixing technology demands lower rotation frequency (0–1500 RPM) and encompass the micro-balloon pumping [134]. With higher centrifugal force, the liquid is pushed into an air chamber inflating a latex micro-balloon. A reciprocating mixing can be performed increasing (inflating the micro-balloon) and decreasing (flattening the micro-balloon) the rotation frequency [134].

Additionally, mixing can be done with gas bubbles integrating external pumps [135] or with gas generation by internal reactions [136]. Figure 9B illustrates the schematics and images from two separated layers of solutions contained in a vented reservoir (top channel) (a), air pushing through the lower channel creates bubbles that promote the mixing (b), thus creating a homogeneous solution in about 100 ms (c) [135]. To avoid the need of external apparatus, Burger et al. [136] implemented a bubble-based mixing by gas produced through a chemical reaction (Figure 9C(a)). The oxygen bubbles are generated in the reaction chamber by the hydrogen peroxide decomposition, which provokes a drag flow through the liquid. A strong buoyancy causes deformation and rupture of these bubbles inducing mixing flows, as represented in (b). In a DNA extraction experiment, this approach allowed a similar performance compared to the "shake-mode" mixing and can be an alternative approach when mixing has to be performed at a fixed rotational frequency.

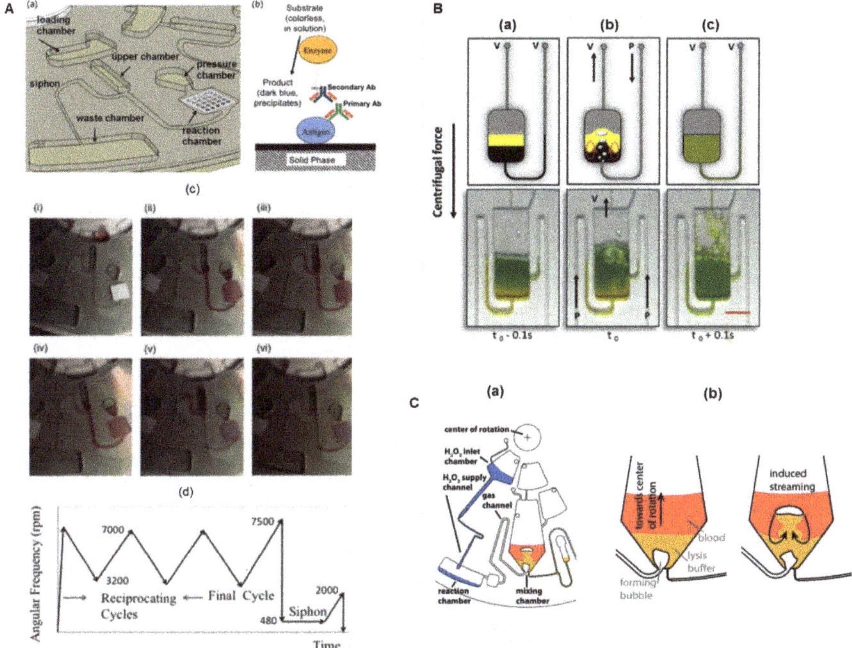

Figure 9. (**A**) Pneumatic pressure flow-based reciprocating mixing. Schematics of (**a**) the fluidic system and (**b**) the antibody capture assay to take place in the reaction chamber; (**c**) images of the system in operation: (i) sample is loaded; (ii) the increase in the rotation frequency promotes the air compression in the pressure chamber; (iii) the decrease in the rotation speed promotes the air expansion that pumps the liquid towards the center; (ii) and (iii) can be repeated in a reciprocated manner. (iv) The increase of rotation to the highest speed (7500 RPM) followed by (v) a decrease to the lowest rotation frequency (480 RPM, see protocol in (**d**)) allow the liquid to prime the siphon valve and (vi) send the liquid to the waste chamber. Reprinted from [132], with permission of AIP Publishing. (**B**) Bubble-based mixing schematics with an external pump: (**a**) from an unmixed state of two dyes, (**b**) under bubbling mixing, to (**c**) a homogeneous solution. Reproduced from [135] with permission from The Royal Society of Chemistry. (**C**) Buoyancy driven bubble mixer set-up. (**a**) Hydrogen peroxide flows into the reaction chamber and the decomposition reaction produces oxygen which is directed through the gas channel to the mixing chamber. (**b**) The mixing chamber with blood and lysis buffer in separated layers (left) and a bubble detaching and ascending through the liquid (right). Reproduced from [136] with permission from The Royal Society of Chemistry.

Mixing and washing steps are crucial for biosensors concerning especially the incubation of the sample with the biorecognition element, which is fundamental for biochemical reactions to occur. Incubation can be performed at a constant low rotation for some assays (400–600 RPM for few minutes [98,137]), while longer periods are necessary for some complex assays such as nucleic acid amplification, e.g., LAMP assays (about 60 min at low or paused rotation [138–141]), NASBA (about 70 min [142]), and RPA (about 15–30 min [143,144]). In ELISA the incubation protocols have been performed even in constant spinning (at 20 Hz for 15 min [85]) or in mixing modes ("shake-mode" mixing from 15 to 30 Hz for 10 min [99] or by magneto-balloon assisted mixing at 900 RPM for 30 min [60]). Incubation protocols should be developed and optimised individually for an assay. In biosensors based on the formation of a complex of magnetic nanoclusters with antibody- [145] or aptamer- [146], the presence of the target triggers the magnetic particles agglutination, which is quantified with an optomagnetic readout. A two-step protocol promoted reliable results with reduced incubation time: the first 1 s incubation step occurs in a strong magnetic field to facilitate clusters formation, followed by the second step of 2 s mixing by shaking to break unspecific bindings and to facilitate the reorientation of the beads for the readout [145,146]. A protocol with 180 cycles of incubation and mixing was established to detect the NS1 Dengue biomarker in serum [145].

After any incubation, the washing step has to guarantee that any impurities, as well as unspecific binders are removed. Washing is usually performed in "shake-mode" mixing with proper buffer solutions, and can be performed more than once if necessary. For each LoaD platform, it is important to select the most suitable approach. "Shake-mode" mixing is the simplest way to achieve good mixing in a passive set-up (no need of external devices), but demands to vary the speed rotation for relatively high frequencies. If this is not possible, external forces (e.g., magnetic or pressure) can be introduced, or bubbles can be generated internally, however, this demands a precious disc area (additional storage reservoirs) and disc area can be quite limited since many lab protocols has to be included in a LoaD biosensor.

4.4. Detection Methodologies

A broad variety of analytical approaches have been used in LoaD biosensors. There is no rule to follow concerning which method should be or should not be used in LoaD platforms, but all of them exhibit advantages and limitations that should be considered. A complete review of detection methods for centrifugal microfluidic platforms is found in [39]. If the analyte is detectable with enough sensitivity and selectivity, the optical colorimetric strategy is the most cost-effective and easy integrated readout method. With technological advances and miniaturisation of detection systems, other approaches raise attention to PoC devices. Special attention to the label-free methods as electrochemical impedance spectroscopy (EIS) and SPR (surface plasmon resonance), which can greatly simplify immunoassays by reducing fabrication-cost and fluid handling. Table 2 lists some detection techniques used in LoaD biosensors comparing the advantages, limitations and solutions, some of them especially thought to adapt the analytical detection set-up to the LoaD platform. Following, this section discusses recent and relevant developments concerning optical and electrochemical detection.

4.4.1. Optical Detection

Optical strategies require three basic components: a light source, a photodetector, and a reaction chemistry, which is analyte concentration-dependent. Particularly to the absorption and fluorescence techniques, considering that the discs can be fabricated in transparent polymers, easy integration of the detection module is possible with the optical transducer placed as a peripheral optical device without the need of electric contacts and wires. Optical strategies for centrifugal microfluidic devices are reviewed previously in [38,39]. Especially for biosensors, optical approaches are widely used because many biological analytes absorb light in a specific wavelength or can be easily marked to do so. In addition, many commercial kits were available for quantitative determination by colorimetric and fluorimetric approaches, which greatly simplifies the R & D stages of LoaD optical biosensors.

Table 2. Detection strategies in LoaD biosensors, advantages, limitations and solutions.

Detection Method	Advantages	Limitations	Solutions
Optical: UV-vis absorbance	Easy integration to LoaD (no contact is needed). Adequate to portable and inexpensive readout device.	Limited to the intrinsic analyte property or demand labelling. Limited sensitivity for thin discs (Abs depends on the light path length).	Colorimetric reagents available for many biochemical reactions. Use of thick discs, or TIR optics to increase path length [147,148].
Fluorescence	High sensitivity. Well-established analytical protocols.	Labelling with fluorescent dyes is needed, leading to complex fluid handling (sandwich assay). Relatively complex and expensive optics sometimes lead to off-disc measurements. Polymer autofluorescence can reduce sensitivity	Availability of highly sensitive and selective fluorescent labels. The use of strobe flash as light source to fluorescence allowed measurements on rotating disc [149]. Select polymers with low autofluorescence (as PDMS) [150]
Surface plasmon resonance (SPR)	AHigh sensitivity. Label-free (simple immunoassay set-up).	Expensive equipment and difficult integration. Need Au or Ag sensing surface integrated to LoaD. Results susceptible to temperature variations.	Simplified optical set-up (but less sensitive) can be implemented to μg/mL IgG detection [87]. Grating based SPR detection uses simplified optics [151].
Electrochemical: Impedance spectroscopy (EIS)	Label-free (simple immunoassay set-up). Independent of sample turbidity or optical path length.	Need electrodes integrated to LoaD. Rotating should be stopped to wires connection and measurement through a potentiostat.	Screen-printed electrodes [152] and gold sputtering [86] were successfully applied to electrode integration on disc. Wireless technology based potentiostat-on-a-disc (PoaD) allows on-disc and rotating disc measurements [153].
Amperometry/Voltammetry	High sensitivity. Fast response. Independent of sample turbidity or optical path length. Adequate to portable and inexpensive readout device. Multianalyte analysis in a single assay (Voltammetry).	Analyte (or co-products) should be electroactive or suspended in electroactive species solution. Need electrodes integrated on LoaD and wires connection to measurement.	Several biochemical compounds are electroactive or produce electroactive products through enzymatic reactions. PoaD can allow on disc measurements.

The Beer–Lambert Law shows the dependence of the light absorption with the absorbing species concentration, which allows a simple approach of sample quantification. A simple set of light emitting diode (LED) with the adequate wavelength emission and a proper photodetector perpendicularly aligned to the disc plane can be used to create a compact and low-cost detection module [154]. The Lambert-Beer Law also states that the absorbance depends on the optical length. Because of that, achieving a reasonable limit of detection in thin discs can be difficult. This can be overcome with thick discs, or by increasing the light path length with an optical beam-guidance set-up based on the total internal reflection (TIR) [147,148].

Various well-established colorimetric lab protocols were successfully implemented in LoaD platforms as loop mediated isothermal amplification (LAMP) and polymerase chain reaction (PCR) as well as the immunoreactions-based fluorescence-linked immuno-sorbent assay (FLISA) and enzyme-linked immuno-sorbent assay (ELISA). LAMP-based LoaD platforms for bacteria [138,139,141] and virus [155–157] detection, as well as PCR for virus [90,95,158] and bacteria [137,159] detection have been published. DNA amplification is a reliable detection method and has existed for decades. The implementation of this technique on a LoaD however is a more difficult task. The LoaD proposed by Li et al. [90] is illustrated in Figure 10, a PoC apparatus for Hepatitis B Virus (HBV) detection which boasts the same diagnostic capabilities as that of a centralised laboratory test. On-board valving allows for serum separation and reagent storage which allow for a fully automated test. A laser diode activates the valves allow for the release of reagents. Centrifugal forces generated by the double rotation axis of the microfluidic platform, which rotate to allow the stationary magnets to displace the internal capture magnetic beads for nucleic extraction. The complex system includes three resistors and thermistors which control the specific temperatures required for the PCR process. Finally, a 470 nm-blue LED served as an excitation source. The fluorescent signal was then detected using two optic fibers when the turntable containing the LoaD was position to the fixed detection region. The only user interaction required is the insertion of 500 μL of whole blood where the systems boasts an LOD of 10^2 copies/mL in just 48 min [90].

As exemplified before, nucleic acid amplification strategies, as LAMP and PCR assays, demand controlled temperature raising. Temperature sensors and heating system can be also integrated into the LoaD platform. Technologies using hot air gun [160] and heater boards [139,155] have been proposed to control temperature profiles. Figure 11A depicts a set-up for a LAMP-based bacterial infection sensing. The heater board with embedded resistive heating elements and thermistors was connected to a printed circuit board (PCB) which allows temperature control and wireless communication with a remote computer [139].

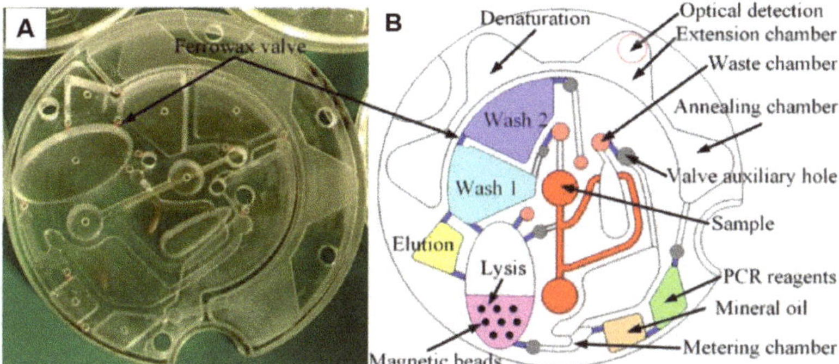

Figure 10. (**A**) The optically transparent disc and (**B**) the schematic of the LoaD architecture for HBV DNA detection in whole blood. Sample is inserted into the centre reservoir of device where pre-stored DNA extraction reagents in wash 1, 2, elution ad lysis buffers and magnetic beads. Ferrowax valves control the release and flow of each reagent to initiate the DNA extraction which then mixes with pre-stored primer and PCR reagents for amplification purposes. Reprinted with permission from [90]. Copyright 2019 American Chemical Society.

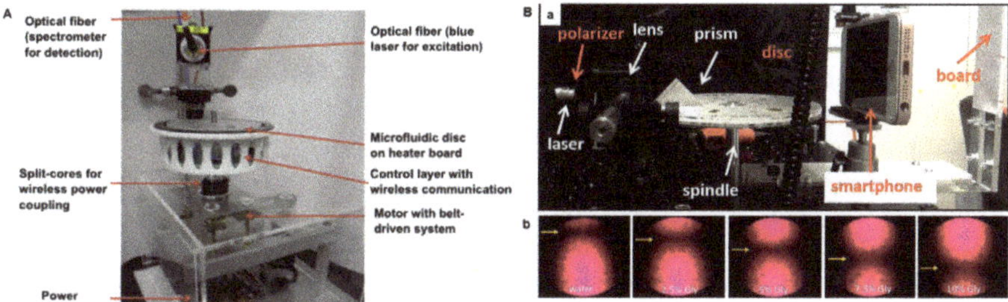

Figure 11. (**A**) LoaD set-up for LAMP assay-based bacterial infection sensor. A belt-driven motor to provide the centrifugal force; power coupling device; wireless data communication module for temperature control and signal detection; light source (blue laser at 488 nm) for the fluorescence excitation; and spectrometer linked to a data acquisition computer. The disc is on an electronic circuit board (PCB) accommodating the heating elements for microfluidic disc to the temperature control. Reproduced from [139] with permission from Elsevier. (**B**) SPR immunosensor set-up: photographs of (**a**) SPR optics integrated to a spin stand, and (**b**) the reflected laser spots from aqueous glycerol pattern solutions (with different refractive indexes) showing the position of the dark lines (caused by the plasmon absorption). Adapted from [87] with permission from Elsevier.

Immunoassays in format of ELISA [99,161] and FLISA [98,162] have also been implemented on LoaD platforms. Conventionally, the detection is made by the absorption (or emission) measurement using a monochromatic light source and an optical sensor, as demonstrated by Thiha and Ibrahim [161] with a PoC sandwich-type ELISA platform for Dengue detection. Oh et al. [163] have applied a similar colorimetric measurement, combined with LAMP, for detection of food-borne pathogens.

Immunoturbidimetry method was used to measure hemoglobin A1c (HbA1c), a glycated hemoglobin recommended by the American Diabetes Association for diabetes diagnosis [164]. The process automated on disc includes the rupture of erythrocytes by a hemolysis reagent, the attaching of the hemoglobin on latex particles, and the specific agglutination of HbA1c, which causes the solution turbidity. The specificity is given by the HbA1c monoclonal antibody added to induce the particles agglutination. A calibration curve reading the absorbance in 660 nm against Hb1Ac concentration was plotted for further quantification of hemoglobin in unknown samples. Fourteen blood samples were tested giving a rapid response (8 min), a good correlation with laboratory results (ion-exchange chromatography), and a standard deviation of ± 0.36% HbA1c (of order with standard laboratory equipment) [164].

Despite ELISA presenting the advantage of the easy colorimetric detection, it requires labeled proteins. Label-free methods as those based on surface plasmon resonance (SPR) simplify the disc design (less reagents and washings) and reduce the production cost (labeled proteins are expensive). SPR biosensing is based on changes in the refractive index very close to the sensor surface (usually gold or silver) caused when the analyte (in solution) binds to the biorecognition element immobilised on the sensor surface [165]. Grating-based [166] and prism-based [87] SPR approaches have been proposed. We have demonstrated an SPR-LoaD platform to simple multi-analyte immunoassay adaptable to detect specific diseases just by changing the capture antibody loaded onto the gold sensor [87]. The gold sensor was attached to the top of the disc immediately before the assay running exempting the need for special storage and transport cares and generalizing the device manufacture. The SPR optics (Kretschmann configuration) were integrated to a spin stand (Figure 11B(a)), and were tested in glycerol solutions to check on the dependence of the SPR angle position with the refractive index of solution. Pictures of the reflected light spot were recorded with a smartphone camera in a dark room (shown in b), and further image analysis provided a plot of reflected light intensity versus incident angle, with the minimum reflectivity (SPR angle) caused by the plasmon absorption. Shifts in the SPR angle were proportional to the antigen concentration when the biorecognition element (specific antibody) was immobilised on the gold sensor. Details can be found in the source [87]. Although the SPR optics can appear to be complex when integrated into LoaD, automated approaches turn SPR a very sensitive optical detection technology for LoaD biosensors. The company Biosurfit (Lisbon, Portugal) already commercialises SPR-based platforms [167].

Raman [168] and surface enhanced Raman scattering (SERS) [169,170] based discs for the detection of organic compounds have been recently published bringing good perspectives for future LoaD biosensors. Because of the expensive and complex detection set-up, these methods are not appropriate to PoC devices at this point. However, certain exigent applications can be worth due to the single-molecule sensitivity [171]. Optical methodologies are far the most explored approach for LoaD platforms due to the cost-effectiveness, the possibility of miniaturisation and personalising of PoC devices, being the closest to fully satisfy the (RE)ASSURED criteria. There are a huge number of publications on optical detection-based LoaD sensors, some of them are summarised in Table 3 to exemplify the variety of detection approaches, analytes, assay types and processes integrated on LoaD biosensors.

Table 3. Summary of some optical detection-based LoaD biosensors. DR: detection range, LOD: limit of detection, WB: whole blood.

Ref	Analyte	Optical Technique (Assay Type)	Total Assay Time	Analytical Data	Comments
[99]	Prostate cancer biomarkers (t-PSA and HER-2)	Absorbance (sandwich and competitive ELISA)	3180 s	LOD: 0.894 ng/mL HER2 and 0.408 ng/mL fPSA	Not tested in real samples.
[164]	Hemoglobin A1c	Absorbance (immunoturbimetry)	8 min	DR: 3.5–12.1 % HbA1c	Input: 1 µL WB 14 samples tested. Standard deviation of 0.36% for each sample (in the order of laboratory equipment)
[172]	Human albumin	Absorbance (sandwich-ELISA)	18 min	LOD: 0.516 ng/mL DR: 0–100 ng/mL	Off-disc absorbance measurement.
[173]	Liver function markers (ALB, TBIL, DBIL, ALP, GGT)	Absorbance (enzymatic assays)	20 min	DR: 0–54.9 µmol/L TBIL, 0–54.9 U/L ALB, 0–48.0 U/L GGT, 0–159 U/L ALP	Input: 150 µL WB All processing integrated in a POC device with wireless communication.
[145]	Dengue NS1 protein	Blu-ray optomagnetic signal (immuno-MNP)	8 min	LOD: 25 ng/mL DR: up to 20,000 ng/mL	Input: 6 µL serum
[146]	Thrombin	Blu-ray optomagnetic signal (immuno-MNP)	15 min 30 s	LOD: 25 pM in PBS	Not tested in real samples.
[110]	C-reactive protein (CRP)	Blu-ray optomagnetic signal (immuno-MNP)	14 min	LOD: 30 µg/mL	Input: 35 µL diluted blood. Dilution performed off-disc.
[91]	Botulinum toxin	Fluorescence (sandwich-immunoassay)	<30 min	LOD: 0.09 pg/mL	Disc placed under microscope
[139]	*Mycobacterium tuberculosis* (TB), *Acinetobacter baumanii* (Ab)	Fluorescence (RTLAMP)	2 h	LOD: 10^3 cfu/mL TB (sputum), 10^2 cfu/mL Ab (blood)	Sputum and WB pre-treatment required. All LAMP processing and detection integrated.
[157]	Avian influenza viruses	Fluorescence (RTLAMP)	70 min	LOD: 25 copies for N3 gene detection	Input: 40 µL WB. All LAMP processing and detection integrated. Wireless communication for temperature control and real time measurement.

Table 3. *Cont.*

[155]	Influenza A virus	Fluorescence (RT-LAMP)	47 min	LOD: 10 copies of RNA	Sample from nasal swabs. 10-fold higher sensitivity than conventional RT-PCR. All LAMP protocols and detection integrated.
[90]	Hepatitis B virus	Fluorescence (PCR)	48 min	LOD: 10^2 copies/mL	Input: 500 µL WB. All PCR protocols and detection integrated.
[174]	Model IgG	Fluorescence (sandwich-ELISA)	62 min	LOD: 20 ng/mL	Disc placed under microscope
[175]	C-reactive protein (CRP) and cardiac troponin I (cTnI)	chemiluminescence (sandwich-ELISA)	30 min	CRP in serum: LOD: 6.0 fM, DR: 8.0 fM–0.8 pM cTnI in WB: LOD: 1.5 pM, DR: 0.4 pM–4.0 nM	Input: 10 µL WB Chemiluminescence measured with a home-built system.
[87]	Model IgG	SPR (direct immunoassay)	<1h	DR: 0–100 µg/mL LOD: 19.8 µg/mL in PBS	Kretschmann configuratio SPR. Developed to input 600 µL WB (not tested in real samples).

4.4.2. Electrochemical Detection

Electrochemical methods are widely used in conventional off-chip sensors because of the fast response, high sensitivity, and relatively low-cost, however, they have been under explored in LoaD devices. This is related to the complexity in integrating the electrochemical signal transducer to the LoaD platform, which demands electrodes included in the disc, and a potentiostat in direct contact with these electrodes. In other words, this usually implicates that the rotation should be stopped to connect the electric contacts to the disc electrodes to perform the measurement. Nwankire et al. [86] implemented gold electrodes functionalised with capture biomolecules where label-free electrochemical detection methods were deployed for measurement. Whole blood was centrifuged on-board, where the cancer cell rich plasma is extracted via a siphon microchannel. Centrifuge-pneumatic DF valves control the realise of reagent for the assay to be carried out. Five identical testing sites exist on the electrochemical-LoaD, allowing for multiple assays being carried out in parallel. The disc was stopped to Electrochemical Impedance Spectroscopy (EIS) measurement for each compartment invidually. The device boasts an 87% capture efficiency over a dynamic range of three orders of magnitude with a lower LOD of 214 cells/mm^2. The lower LOD equates to just 2% of the total working electrode surface [86]. The detection method of label-free EIS allowed for rapid cancer cell detection. The change in the electrical interfacial properties of the surface due to biological binding events is detected. For label-free detection, the changes in the dielectric constant, resistance and capacitance on the surface due to the capture of the target molecule is measured. The change in the impedance would be an indication of whether an analyte has been captured or not, and depending on the change, whether it is a high or low concentration of the target analyte.

Electrochemical biosensors of this nature have attracted major interest for label-free analysis as they don't require as many assay steps as their labelled counterpart for signal acquisition. This form of biosensing attains sensitivity and simplicity which makes them a reliable, quantifiable diagnostic tool for whole cell capture and detection [176]. Opting for whole cell detection rather than DNA detection also reduces the cost and time of detection, as no cell lysis or additional reagents are necessary. The use of sputter-coated electrodes allows for the platform to be disposed of after a single-use and more cost effective, making this type of PoC device economical. These results give a true insight into the huge potential that exists for electrochemical-LoaDs as they are an efficient prognostic devices where minimal sample preparation is required and can be further developed for clinical applications [86]. The relative changes in impedance (difference between values of the anti-EpCAM coated gold electrodes before and after interaction with different concentrations of SKOV3 cells) varies linearly with the captured cell number [86].

To overcome the drawback concerning the integration of the detection system, recent works have focused on the electrochemical modules to allow measurements during the disc rotation [25,153]. This opening up opportunities of real-time monitoring of bioreactions on disc [25,153]. Andreasen et al. [25] developed a low-noise component (electrical slip-ring) to on-disc electrochemical measurements. Cyclic voltammetry tests performed from 0–600 RPM indicated only a small perturbation of the peak currents with the spin rate. Recently, Rajendran et al. [153] demonstrated a modular lightweight (127 g) and wireless potentiostat-on-a-disc (PoD) able to perform square-wave voltammetry (SWV) and amperometry, controlled by a custom-made software. The current resolution of 200 pA was in agreement with commercial potentiostats. During the experiment, data was transmitted via Bluetooth to a Windows PC, plotted live and displayed in a custom-developed LabVIEW program (Figure 12A). The PoD detection unit was composed by a Qi-based wireless power supply, a core circuit platform and a module (shield) (an exploded view in Figure 12B). Figure 12C(a) presents the cyclic voltammograms (CV) of 1 mM potassium ferricyanide (FiC) collected in stationary-mode using the PoD and a commercial potentiostat and (b) CVs collected in different spinning rates by the PoD. At 0 and 1 Hz, the reaction was diffusion-limited while reaction-limited (due to the mass transfer) at 2 Hz and above [153]. The comparison between PoD and the commercial equipment was extended to SWV: (c) voltammograms of 10 mM ascorbic acid and (d) the respective calibration curve. Both cyclic voltammetry and SWV measurements obtained by the PoD were comparable to the commercial system. The dependence of the amperometric response with the spin rate was tested in a multichannel shield, shown in (e), with the current response for 500 µM FiC increasing with the rotation frequency, as observed in current *versus* FiC concentration plot, shown in (f). Refer to the source for details [153].

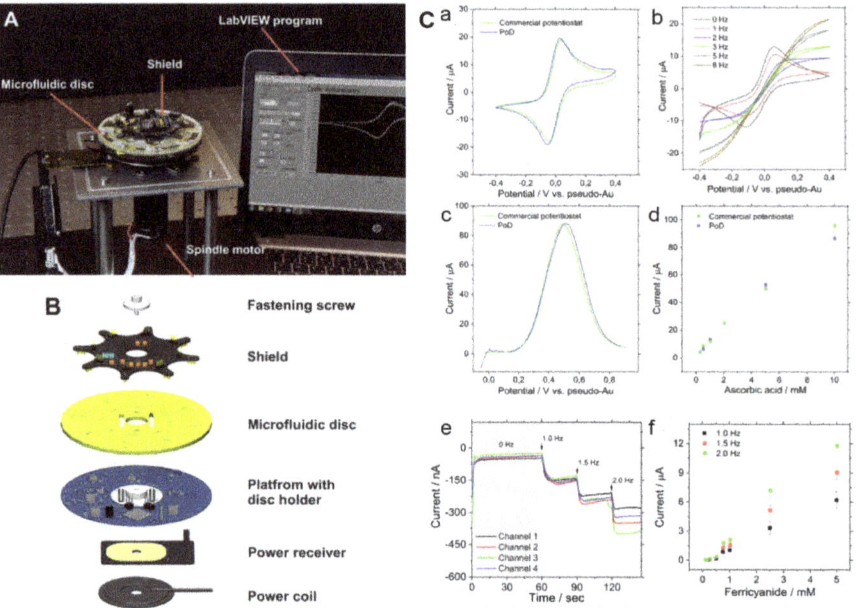

Figure 12. (**A**) Photograph of the experimental set-up with the spindle motor, LoaD platform, PoD and computer to control the software interface. (**B**) Exploded view of the PoD and LoaD device. (**C**) PoD results: (**a**) CVs of 1 mM FiC (Au working electrode with area of 50 mm^2, scan rate of 50 mV/s comparing PoD and commercial potentiostat results, and (**b**) CVs collected in various rotation speeds at scan-rate of 100 mV/s; (**c**) SWVs of 10 mM ascorbic acid (carbon working electrode with 50 mm^2, potential step 0.004 V, frequency of 10 Hz and amplitude of 0.025 V), and (**d**) calibration curve for ascorbic acid concentrations. (**e**) Amperometric response for 500 µM FiC (pH 7.4 PBS as supporting electrolyte, applied potential of −0.4 V *vs* pseudo-Au reference electrode; working electrode area: 0.69 mm^2; Au counter-electrode) in different rotation speeds, and (**f**) the normalised current changes recorded for different FiC concentrations during rotation at 1.0, 1.5, and 2.0 Hz (n = 3). Adapted with permission from [153]. Copyright 2019 American Chemical Society.

Sanger et al. [177] report a centrifugal platform capable of capturing and sensing pathogens. The robust LoaD contains an integrated sample pre-treatment system where SWV are used for cell free detection of a secondary metabolite, p-Coumaric acid (pCHA), a biomolecule produced by genetically modified strains of E-Coli bacteria. A common attribute of LoaD platforms is their ability to contain multiple, identical testing sites. A total of eight regions are available for individual biosensing, where a 0.2 µm filter membrane system exists in each section for sample filtration. This filtration and separation occur under centrifugal conditions for 5 min at approximately 12 Hz. The supernatant is metered and then displaced into the detection chamber. Unlike other electrochemical-LoaDs, the sensors are patterned onto an acrylic base of the platform using e-bean evaporation. Contact points connecting both potentiostat and sensor exist through the depth of the LoaD to establish a connection through a slip-ring. Measurements are obtained under static conditions and showed very promising results for the detection and quantification of the two E. coli pCHA metabolites compared to results obtained from other detection methods such as HPLC [152].

This group also reported a novel, on-board pre-treatment based on supported liquid membrane (SLM) technology. This extraction method is used to purify and enrich analytes directly from complex matrices such urine and saliva [23,49]. Using the SLM to facilitate on-disc separation, the main hindering compound, Tyrosine, is subsequently removed and the pCHA is enriched. The platform enables multiple measuremtns to be obtained at certain stages of the production process when the target analyte is quite low (typically 100 µM). The main biosensing technique deployed were CV and SWV measurements. As the device is deemed as low-cost it was aimed to be a single use device, however it has been shown that the device can be re-used using typical electrode cycling to clean the surfaces. This work poses a significant insight into the exciting advantages and potential of on-disc sample pre-treatment and electrical detection for analytes in a compact, portable testing [152].

Amperometry is another electrochemical technique where the current response due to a redox on the surface of the biosensor is measured over a period of time at a constant potential [178,179]. It is often used as a detection technique when coupled with electrochemical-LoaD platforms. The potential is maintained at the working electrode via the reference electrode where the current is recorded due to oxidation or reduction of electroactive species. In cases where the analyte is not electrochemically active, the target cannot be measured directly but indirect approaches can be used. Screen-printed carbon electrodes (SPCE) which were modified using graphene-polyaniline nanocomposites were incorporated onto a LoaD to enhance electrochemical detection of glucose. The enzyme and glucose solutions, loaded into separate chambers, mix in a serpentine channel under centrifugal conditions creating the hydrogen peroxide through the enzymatic reaction in under 8 min. The electrochemically active hydrogen peroxide was quantified to deduce the amount of glucose present in the sample, with a reported LOD of 0.29 mM [152]. Common interfering compounds ascorbic and uric acid were tested where the sensor remained specific towards glucose. This is a critical attribute of a biosensor, as it is important to determine to working capability of the sensor for real applications.

Li et al. [180] created a whole blood (WB) analysis system with a primary functionality of performing basic metabolic testing. Different concentrations of glucose, uric and lactate acids in WB spiked samples were analysed within minutes [180]. WB is a very difficult medium to work with, as various blood components non-specifically bind to sensor surfaces. Figure 13(A) depicts the design of one of the testing sections comprising a microfluidic layer, an insulating layer an a biosensing layer. The system set-up contains a nano-porous Au plated Si electrode. Carbon nanotubes on the working electrode increases the electrode surface area which increases the rate of immobilisation of Prussian blue (catalyst) and the analytes. Sample sizes of just 16 µL was required to perform each test, where each PDMS device contained eight individual testing sites. Amperometry allowed for rapid detection and LODs of 0.3 mM of glucose, 0.1–0.2 mM of uric acid and 0.7–1.5 mM. Figure 13(B) illustrates (a) the WB sample loaded. The centrifugation of the WB samples separates the blood components and plasma for analyte capture and electrochemical detection, as illustrated in (b). To further verify the capabilities of the device, a comparison study was carried out using a commercial colorimetric assay, where results indicated deviations for glucose, lactate and uric acid of 4.2%, 5.2% and 6.0%, respectively, between both methods [180].

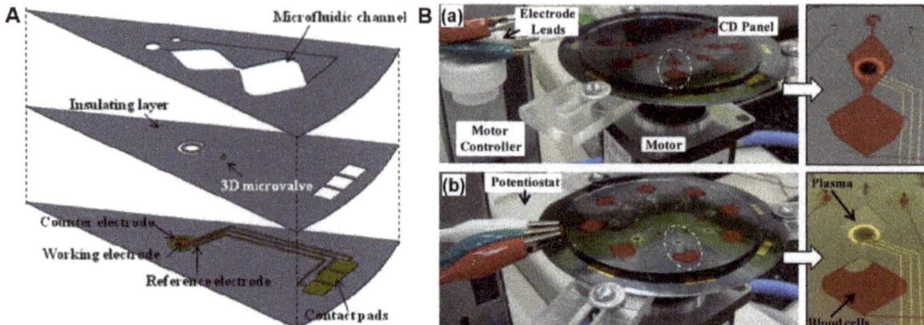

Figure 13. (**A**) Conceptual design of one of eight testing sites displaying each consecutive layer of the PDMS electrochemical-LoaD: a microfluidic layer, an insulating layer and a biosensing layer. (**B**) Electrochemical-LoaD whole blood analysis: (**a**) sample loaded ready for rotation, and (**b**) sample centrifuged and the plasma isolated for measurement of electrochemical signal. Reproduced from [180] with permission from The Royal Society of Chemistry.

Amperometry on a LoaD was again reported by Kim et al. [23] where the cardiac marker, C-reactive protein (CRP), was detected in 20 min. The on-board ELISA assay detected using gold electrodes, fabricated using E-beam lithography. 300 Å of chromium was deposited acting as an adhesion layer, followed by deposition of 3000 Å of gold to create the desired electrode pattern for the electrical connection [181]. Ferrowax valves allows for on-board reagent storage which are actuated using a mobile laser under static conditions [37]. The CRP LOD for this system was 4.9 pg/mL, a 17-fold improvement in quantification by optical density (OD) which is the standard ELISA detection method. This device adds to the very small research bracket of microfluidic-electrochemical crossovers for successful detection for very low concentrations of biomarkers.

5. Biosensing on Existing and Emerging Commercial Point-of-Care Lab-on-a-Disc Devices

Among the first companies to offer a LoaD platform on the market was Abaxis with the Piccolo Xpress [182]. In 1995 they launched the Primary Health Panel (a nine-test reagent disc), and the General Health Panel (a 12-test reagent disc). The Liver Panel Plus arrived the next year. Nowadays, Abaxis offers fast blood analysis (about 12 min) with 31 tests across 16 complete chemistry panels [182]. Thenceforth, other assays became available, as the PCR-based devices from Revogene Platform by Meridian Bioscience [183] (tests for gastrointestinal conditions, pediatric and neonatal conditions, respiratory conditions and enterobacteria infection), and the Simplexa assays to be run in 3M Integrated Cycler [184] (real-time PCR for quantitative and multi-analyte detection of virus infections, as well as to develop extraction and amplification assays using the kits the company offers in its catalog) .

Strohmeier et al. [34] presented a complete list of companies selling LoaD platforms until 2015 in a review paper. On this occasion, Hahn-Schickard [185] was cited as in the development stage. Today, this German company operates as an R & D service provider, developing solutions in microsystem technology, with a wide portfolio of services. Recently, Hahn-Schickard and related company Spindiag GmbH jointly received an investment of 6 million Euros from the German government to establish a SARS-CoV-2 virus rapid test to address COVID-19 [186] based upon their existing platforms. Spindiag [187] is a medical technology specialised start-up founded in 2016 as a spin-off of the Hahn-Schickard. In May 2020, Spindiag announced an additional 16.3 million Euros investment to prepare market launch of the Rhonda mini-lab for SARS-CoV-2 virus. The Rhonda system is a PCR-based device and will provide results in 30–40 min with minimal contact between the operator and sample. Swab samples from the nose and throat are directly inserted into the cartridge, which contains all the required reagents. The system is on an analytical test and it is expected to receive market approval in Germany and European Union during the third quarter 2020 (information from Spindiag website [187]).

Immunoassays often require multiple steps such as sample incubation, washing and elution steps, blood separation and a final detection output. Gyros ABTM has proven commercial success with a first-generation centrifugal testing platform for immunoassay applications. The platform uses hydrophobic regions to regulate the sequential flow inside the injection moulded device which permits on-board metering and aliquoting of reagents. One device provided by the company, Gyrolab Bioaffy CD [188], has up to 112 channels which allows for high throughput testing of the sample where the users own specific biomarker may be detected. The device contains a series of metering structures, which evenly distributes the capture reagent. Streptavidin coated beads in each region are incubated with the capture biomolecules, where the excess reagent is displaced into another chamber by increasing the spin frequency of the platform. Samples, of 200 nL defined volume, enter the incubation area by capillary action where the analyte is captured. A secondary detection reagent is then passed through the capture column, creating a sandwich assay. This nanolitre immunoassay significantly reduces sample and reagent usage where fluorescence detection from the assay is reportedly obtained within the hour. The unique selling point of the device is that specific immunoassay can be designed for implementation on the device based on the required dynamic range and analyte concentration where the company itself will provide support whilst designing the LoaD assay. Ghosh et al. [189] reported using the commercially available Bioaffy 1000 CD to measure protein drugs capable of neutralises vascular endothelial growth factor to save patient vision with retinal vascular diseases. Cynomolgus ocular final concentrations were determined using an Alexafluor labelling kit and where visualisation and measurements were achieved using laser-induced fluorescent detection at a 647 nm wavelength [189]. Many advantages are offered by this device regarding assay volume sizes being realised with this platform, there are many clear drawbacks with the concept. The company does not provide with ready to use devices, where all biorecognition molecule pairs for the sandwich assay require specific user knowledge. As the assay detects biomolecules, some form of sample preparation is required which cannot be performed on the LoaD [34,190].

The LabGEO IB10 device is commercially available centrifugal platform distributed by Samsung. The device is an integrated system that combines a microfluidic platform and an analyser for cardiac biomarker detection from whole blood. This LoaD only requires 500 µL of a whole blood sample from the user, where all other reagents are pre-loaded. Proteins such CK-MB, Troponin and Myoglobin are extracted from whole blood and tested within 20 min where an optical density ranging over 10 specific wavelengths are obtained by the accompanying centrifuge analyser platform [191]. The platform contains freeze-dried reagents for blood chemistry analysis. The fluid flow is controlled using laser actuated ferrowax microvalves to activate specific reagent steps for the assay. All samples processing is carried out automatically. Other commercial LoaDs have combined both testing and optical density analysis conjoined system [154] however, Samsung appear to be the only competitors on the current market which have been able to provide analysis using whole blood analyte extraction on a PoC system.

Another example of an optical detection system is the ViroTrack® cartridge produced by BluSense Diagnostics (Denmark). Although the chip is not a LoaD by definition, it is a segment which operates using centrifugal microfluidic technology. The ViroTrack® is a single use cartridge which provides results with just one drop of blood in 10 min. The assay technology contained within the chip uses magnetic nanoparticles (MNPs) coated with capture biomolecules that will react and capture the specific target. These MNPs are released which bind with the target biomolecule within the blood. After the incubation period, a strong magnetic field aggregates with MNPs where the density of these are subsequently measured by the BluBox instrument to produce semi-quantitative results. Various versions of the chip are available to test for Dengue, Zika and Chikungunya, Covid-19 viruses [192,193].

Although there are already some LoaD platforms on the market, there is still room for extensive research into new devices. Researchers from University of Freiburg developed a disposable ready-to-use microfluidic cartridges (termed GeneSlices) for PCR assays [194,195]. Four GeneSlice cartridges were fit in a custom made rotor replacing the standard rotor in a commercial Rotor-Gene (Qiagen, Germany) [194,195]. In collaboration with Hahn-Schickard, they created an automated PCR-based disc for bacterial pathogen detection to run on a portable LabDisk-player [137] with a minimal manual step since the reagents and buffers were pre-stored in stick-packs [94]. Pathogens detection via LAMP nucleic acid amplification have been also developed [196,197]. An extreme PoC device for malaria field detection was based on a real-time LAMP assay in a compact analyzer and a disposable microfluidic compact disc. The device was capable to process 4 samples simultaneously within 50 min, with a material cost ~$1/test, and achieving a LOD of 0.5 parasites/µL whole blood (sufficient to detect asymptomatic parasite carriers) [196].

One of the most promising LoaD devices is in the field of circulating tumor cells (CTC). In 2014, Professor Cho's group from Ulsan National Institute of Science and Technology (UNIST) suggested a LoaD platform for CTC isolation by mechanical filtration. The size-selective CTC detection system used a commercially available track-etched polycarbonate membrane with 8 µm-pore-size, specially chosen based on the size of the CTC. The entire process of CTC isolation, staining and detection was integrated on disc by a programmable operating system. When compared to the commercial ScreenCell, the caption efficiency for the MCF-7 cells was slightly lower (56 $\pm$ 2% compared to 69 $\pm$ 6%) (n = 3), however, the contamination by white blood cells was 20-fold lower with the disc system [198]. In 2017, they suggested the fluid-assisted separation technology (FAST) for CTC isolation using the same filter but now with the pores filled with a stably held liquid during the entire filtration process, achieving an impressive recovery rate of 95.9 $\pm$ 3.1% [199]. While in a non-FAST mode, the filtration left the majority of particles located in the outer rim of the filter (only a small part of the filter is used), the membrane fully wet before and during the filtration process allows the liquid transport at a pressure lower than the capillary pressure, leading to a uniform filtration throughout the entire membrane [116]. The FAST device was successfully validated in a pre-clinical test in unprocessed whole blood of 40 patients with non-small cell lung cancer [116]. FAST disc was patented, licensed to Clinomics [200] (Ulsan, Korea). A pilot study using the FAST device was performed in whole blood of 13 patients with ovarian cancer, for isolation and enumeration of CTC, and to assess the correlations among CTC, cancer antigen-125 levels and the clinical course of the disease [201]. The FAST disc presents great potential to provide minimally invasive way to detect metastasis and for monitoring treatment response in patients.

Through technological advances, the current wireless technology and Bluetooth connectivity can allow the complete automation of processes on disc. Besides the control of spinning protocols, important experimental parameters, as valve action and temperature control [202], as well as real-time data acquisition [141,153,154,203–205] can be performed by Bluetooth connection. To avoid stopping the disc and/or off-disc measurements, an electrified-LoaD (eLoaD) platform [203,204] was created to allow continuous measurement while the disc is spinning with data transmission via Bluetooth, which has special importance for time-dependent reactions and biochemical reaction kinetics. The platform is assembled on a modular set-up, with an interchangeable and non-disposable "application disc" specifically designed for the intended application, sandwiched between the LoaD microfluidic disc (top) and the eLoaD platform (bottom). Authors highlighted the versatility by adapting the "application disc" to the intended protocol, for instance, manufacturing an "application disc" for optical detection by housing proper set of LEDs and photodetectors, or by housing a set of temperature sensors and heaters to automate the temperature control, etc. [204]. The wireless power is transferred during the disc rotation from a 5 W-Qi Standard transmitter, commonly used to charge smartphones. An Arduino microcontroller, a Bluetooth communication module and an SD-card module are settled in the platform [204]. The portability and the easy control by Tablets and Smartphones can be a key feature for extreme PoC devices, and for daily diagnostics for a rapid data transfer to health care professionals by remote consultation.

Technological advances in materials processing and digital innovations (mobile-health and internet of things) have contributed to create diagnostic tests to satisfy (or nearly satisfy) the REASSURED criteria for PoC devices [31].

- **R**eal-time connectivity: Wireless Bluetooth connectivity in LoaD has allowed both operation control and result visualisation [26,154,161,202,204]. It can be implemented to allow that the results are rapidly provided to the patient as soon as the test is completed.
- **E**ase of specimen collection and **E**nvironmental friendliness: It is preferable that tests with blood utilise small volumes to allow self-collection using finger stick kits. The blood processing is easily integrated on disc rather than any other microfluidic separation strategy. Some tests can be less invasive using swabs for sample collection (vaginal, nasal, oral) allied to PCR detection [32]. The environmental point-of-view draws attention to the cumulative effect and environmental risks concerning the disposal of contaminated discs. Although often neglected in the published works, it is time to start considering the recyclability of materials in the manufacture of the biosensors. The viability of small interchangeable cartridges to contaminated solutions could eventually allow the correct disposal of the contaminated residues (incineration), while a great part of the disc can be directed for recycling.
- **A**ffordable tests can be achieved thanks to the relatively cheap and easy processing of polymers and possible scale production. Certainly, the most expensive parcel of the total cost of a biosensor is related to the biochemical reagents, which also implicate in additional costs in special storage and transport conditions.
- **S**ensitive and **S**pecific LoaD biosensors have been reported for various analytes. The first is provided by the combination of sensitive detection methods and well-done sample processing (reduced background noise) while the second is provided by the biospecificity of the biorecognition element (enzyme–substrate or antigen–antibody specificity and nucleic acid sequences complementarity).
- **U**ser friendliness: LoaD biosensors can be implemented to satisfy this criterion since all the disc processing and data collection and analysis can be automated, starting with a single "play" bottom push.
- **R**apid and **R**obust tests: LoaD can address these criteria. For example, LoaD immunoassays were completed with significantly reduced assay time compared to common benchtop ELISA (blood separation time reduced to about 8 min while sample incubation is successfully achieved in 10 min [87,99]). Concerning robustness, some disc designs allow the placement of the biosensing surface to the disc just immediately before the test, facilitating disc fabrication and avoiding special transport and storage conditions [87].
- **E**quipment-free and **D**eliverable: Inexpensive discs can be manufactured and easily delivered, the need for equipment including programable rotation motor and detection system still keep the LoaD technologies restricted to laboratory walls. For this reason, LoaD diagnostic tests in the market still target the diagnostic laboratories and are not available as self and home tests.

6. Future Perspective of Biosensing Platforms

Biosensing LoaDs performing bio-assays often requires complex sample processing with an increased level of integration, making them robust and effective tools in PoC testing. Advancements have been made in LoaD manufacturing and fabrication strategies in the past decade, however significant challenges are prevalent for the successful integration of microfluidic PoC systems into routine clinical diagnosis [206]. One of the main challenges for centrifugal microfluidics to become commercially desirable is extending the shelf life of the PoC device. For biosensing and assay performance, multiple reagents are necessary for testing. For these platform to remain highly attractive products on the market, the LoaDs must be semi/fully automated. This attribute is dependent on multiple factors, including;

1. On-board reagent storage: The requirement of assay reagents has limited the development of clinical microfluidic devices, however advancements have been made in recent years to combat this issue. Reagents can be integrated and stored within LoaDs in a variety of ways. Dry reagent storage in can be achieved through inkjet printing and lyophilisation [207,208]. Dehydration of the compounds often impedes the degradation of the reagents and therefore increases the shelf life of the devices in varying environments. This in turn allows for mass manufacturing and storage of the platforms. This will allow for already existing biological assays to be implemented onto the centrifugal platform as a ready-to-use diagnostic tool [209].
2. Reagent actuation: Nucleic acid analysis is a multi-step assay, consisting of cell lysis, nucleic acid enrichment and amplification, and signal readout. The integration of these types of assays calls for clever microfluidic engineering for sequential reagent actuation which can be offered by laser, electro, photo and centrifugal induced pressure triggering valves, siphons and graphene oxide valves for fluid specific routing of liquids [210–212]. This allows the devices to become "ready-to-use" tests within a clinical setting. The majority of these options have existed for some time, however, within the last decade, research groups are now putting an emphasis on inventing specific μTAS [213]. Interdisciplinary groups are innovating LoaDs with complementary detection systems built around the architecture of the diagnostics platform. Previously discussed literature has proven the value of this type of integration i.e., centrifuges with built in heating blocks for rapid PCR assays [214], optics for optical density and fluorescence detection [90] and modular wireless potentiostat-on-a-disc for electrochemical detection [153]. These systems have only become evident in recent years and are becoming increasingly popular within literature. This trend is expected to grow in the next coming years. BluSense Diagnostics have already reached commercial success by offering a centrifugal style chip requiring one droplet of blood to carry out the disease diagnostics tests on their custom centrifuge analyser system [215].
3. Nanotechnology integration: Nanotechnology fabrication techniques have also significantly advanced in recent years, which are now being implemented into PoC devices for more specific detection of analytes. Deployment of nanoparticles coated with specific antibodies/antigens, magneto-nanoparticles, micropillar structures, Au cavities and AuNPs are deployed within centrifugal sensing platforms are reporting greater sensitivity which rivals that of gold-standard assays currently used [216,217]. The nanostructures deployed during testing are more specific towards the analyte and therefore reduces the LOD. As well as the greater sensitivity offered by nanostructures they also assist in overcoming standard microfluidic phenomena. Bioassays require adequate mixing of various phases (liquid, silica beads, etc.) which is difficult to achieve in microfluidic platforms due to the low Reynolds number of liquids in reservoirs and channels. Materials and techniques are now being explored thanks to developments in nanotechnology fabrication. Micropillars and other structures created using lithographic and additive manufacture techniques can disrupt laminar flow in channels, creating the desired mixing motion necessary for these assays [218]. This will need to be explored in the future, however presently these are considered complex manufacturing processes and would be difficult to mass manufacture for commercial success.
4. Additive Manufacturing: The availability of high resolution 3D printing techniques to researchers may facilitate the fabrication and mass production of particularly complicated LoaD design with minimal variation between batch production, making assay performance repeatable and reliable [219,220]. Advances with materials and technologies paired with clever design and fabrication of LoaDs may make laborious clinical application of microfluidics achievable in the near future.

7. Final Remarks

In search of complementarity between microfluidics and biosensing, the centrifugal microfluidic platform has emerged as an preeminent platform of significant interest to academic researchers and commercial enterprises. The key advantage is the capability of the LoaD to enable transfer of LUOs (pipetting, metering, mixing etc.) directly from the conventional lab "wet bench" to "on disc" while requiring minimial protocol changes or optimisation. This ease of automation results in faster and more robust assay development compared with other emerging PoC platforms, such as paper microfluidics [221], where advantages such as greater deployability (for example not requiring a spindle motor to function) are counterbalanced by a greater upfront burden of assay development or biosensor optimisation. Furthermore, the LoaD platform is continuously evolving and has been shown to be a platform which can be adapted to leverage the latest emerging technologies (such as the Arduino controllers and wireless mobile-phone charging technology described above). Every year new LoaD automation and biosensing technologies have been developed to demonstrate sample-to-answer PoC biodetection with minimal user interaction. Even with devices already commercialised, there is plenty of space for researchers to develop the LoaD with new technologies to address new challenges. We hope this review paper has provided researchers interested in working in this field a comprehensive overview of the recent advances on LoaD biosensors.

Author Contributions: Writing—original draft preparation, C.M.M., E.C., and D.J.K.; writing—review and editing, C.M.M., E.C., and D.J.K. All authors have read and agreed to the published version of the manuscript.

Funding: Sao Paulo Research Foundation grants 2014/15093-7 and 2015/16311-0 and partly funded by Enterprise Ireland under Grant No. CF/2019/1080.

Conflicts of Interest: The authors declare no conflict of interest.

Abbreviations

The following abbreviations are used in this manuscript:

COP	Cyclic olefin polymer
CRP	C-Reactive Protein
CTC	Circulating Tumor Cells
CV	Cyclic Voltammogram
DF	Dissolvable Film
DGM	Density Gradient Medium
DNA	Deoxyribonucleic Acid
eLoaD	electrified-Lab-on-a-Disc
EDC	N-(3-dimethy-laminopropyl)-N -ethylcarbodiimide hydrochloride
EIS	Electrochemical Impedance Spectroscopy
ELISA	Enzyme-Linked Immunosorbent Assay
FAST	Fluid-Assisted Separation Technology
FiC	Potassium ferrycyanide
FLISA	Fluorescence-Linked Immunosorbent Assay
HBV	Hepatitus B Virus
ID	Individually Addressable Diaphragm
IVD	In Vitro Diagnostics
LAMP	Loop-mediated Isothermal Amplification
LED	Light Emitting Diode
LoaD	Lab-on-a-Disc
LOD	Limit of Detection
LUO	Laboratory Unit Operations
MNPs	Magnetic Nanoparticles
µTAS	micro-Total Analysis System
NASBA	Nucleic Acid Sequence-Based Amplification
NHS	N-hydroxyuccinimida
OD	Optical Density
PCR	Polymerase Chain Reaction
PEI	Poly(ethylene imine)

PMMA	Poly(methylmethacrylate)
PoC	Point-of-Care
PoD	Potentiostat-on-a-Disc
RBC	Red Blood Cells
RPA	Recombinase Polymerase Amplification
SLM	Supported Liquid Membrane
SPCR	Screenprinted Carbon Electrode
SPR	Surface Plasmon Resonance
SWV	Square-wave voltammetry
TOS	Triangle Obstacle Structure
WB	Whole blood

References

1. Nagel, B.; Dellweg, H.; Gierasch, L. Glossary for chemists of terms used in biotechnology (IUPAC Recommendations 1992). *Pure Appl. Chem.* **1992**, *64*, 143–168. [CrossRef]
2. Reddy, S.M.; Paixão, T.R.L.C. *Materials for Chemical Sensing*; Springer: Berlin/Heidelberg, Germany, 2017; p. 268. [CrossRef]
3. Whitesides, G.M. The origins and the future of microfluidics. *Nature* **2006**, *442*, 368–373. [CrossRef] [PubMed]
4. Tian, W.C.; Finehout, E. Introduction to microfluidics. In *Microfluidics for Biological Applications*; Springer: Berlin/Heidelberg, Germany, 2008; pp. 1–34.
5. Madou, M.; Zoval, J.; Jia, G.; Kido, H.; Kim, J.; Kim, N. Lab on a CD. *Annu. Rev. Biomed. Eng.* **2006**, *8*, 601–628. [CrossRef] [PubMed]
6. Analytics, C. Web of Science. Available online: www.webofknowledge.com (accessed on 18 May 2020).
7. Kung, Y.C.; Huang, K.W.; Chong, W.; Chiou, P.Y. Tunnel Dielectrophoresis for Tunable, Single-Stream Cell Focusing in Physiological Buffers in High-Speed Microfluidic Flows. *Small* **2016**, *12*, 4343–4348. [CrossRef] [PubMed]
8. Chuang, C.H.; Chiang, Y.Y. Bio-O-Pump: A novel portable microfluidic device driven by osmotic pressure. *Sens. Actuators B Chem.* **2019**, *284*, 736–743. [CrossRef]
9. Sotoudegan, M.S.; Mohd, O.; Ligler, F.S.; Walker, G.M. Paper-based passive pumps to generate controllable whole blood flow through microfluidic devices. *Lab Chip* **2019**, *19*, 3787–3795. [CrossRef]
10. Glawdel, T.; Ren, C.L. Global network design for robust operation of microfluidic droplet generators with pressure-driven flow. *Microfluid. Nanofluid.* **2012**, *13*, 469–480. [CrossRef]
11. Laschi, S.; Miranda-Castro, R.; González-Fernández, E.; Palchetti, I.; Reymond, F.; Rossier, J.S.; Marrazza, G. A new gravity-driven microfluidic-based electrochemical assay coupled to magnetic beads for nucleic acid detection. *Electrophoresis* **2010**, *31*, 3727–3736. [CrossRef]
12. De Groot, T.; Veserat, K.; Berthier, E.; Beebe, D.; Theberge, A. Surface-tension driven open microfluidic platform for hanging droplet culture. *Lab Chip* **2016**, *16*, 334–344. [CrossRef] [PubMed]
13. Xu, L.; Lee, H.; Jetta, D.; Oh, K.W. Vacuum-driven power-free microfluidics utilizing the gas solubility or permeability of polydimethylsiloxane (PDMS). *Lab Chip* **2015**, *15*, 3962–3979. [CrossRef] [PubMed]
14. Kung, Y.C.; Huang, K.W.; Fan, Y.J.; Chiou, P.Y. Fabrication of 3D high aspect ratio PDMS microfluidic networks with a hybrid stamp. *Lab Chip* **2015**, *15*, 1861–1868. [CrossRef]
15. Luo, Y.; Qin, J.; Lin, B. Methods for pumping fluids on biomedical lab-on-a-chip. *Front. Biosci.* **2009**, *14*, 3913–3924. [CrossRef] [PubMed]
16. Ahmed, D.; Mao, X.; Juluri, B.K.; Huang, T.J. A fast microfluidic mixer based on acoustically driven sidewall-trapped microbubbles. *Microfluid. Nanofluid.* **2009**, *7*, 727. [CrossRef]
17. Ter Schiphorst, J.; Saez, J.; Diamond, D.; Benito-Lopez, F.; Schenning, A.P. Light-responsive polymers for microfluidic applications. *Lab Chip* **2018**, *18*, 699–709. [CrossRef] [PubMed]
18. Fan, Y.; Wu, Y.; Chen, Y.; Kung, Y.C.; Wu, T.; Huang, K.; Sheen, H.J.; Chiou, P.Y. Three dimensional microfluidics with embedded microball lenses for parallel and high throughput multicolor fluorescence detection. *Biomicrofluidics* **2013**, *7*, 044121. [CrossRef] [PubMed]

19. Gilmore, J.; Islam, M.; Martinez-Duarte, R. Challenges in the use of compact disc-based centrifugal microfluidics for healthcare diagnostics at the extreme point of care. *Micromachines* **2016**, *7*, 52. [CrossRef]
20. Huang, Y.; Wang, J.; Zhang, M.; Zhu, M.; Wang, M.; Sun, Y.; Gu, H.; Cao, J.; Li, X.; Zhang, S.; et al. Matrix-assisted laser desorption/ionization time-of-flight mass spectrometry for rapid identification of fungal rhinosinusitis pathogens. *J. Med. Microbiol.* **2017**, *66*, 328–333. [CrossRef]
21. Kinahan, D.J.; Kearney, S.M.; Kilcawley, N.A.; Early, P.L.; Glynn, M.T.; Ducree, J. Density-gradient mediated band extraction of leukocytes from whole blood using centrifugo-pneumatic siphon valving on centrifugal microfluidic discs. *PLoS ONE* **2016**, *11*, e0155545. [CrossRef]
22. Van de Groep, K.; Bos, M.P.; Varkila, M.R.; Savelkoul, P.H.; Ong, D.S.; Derde, L.P.; Juffermans, N.P.; van der Poll, T.; Bonten, M.J.; Cremer, O.L.; et al. Moderate positive predictive value of a multiplex real-time PCR on whole blood for pathogen detection in critically ill patients with sepsis. *Eur. J. Clin. Microbiol. Infect. Dis.* **2019**, *38*, 1829–1836. [CrossRef]
23. Kim, T.H.; Abi-Samra, K.; Sunkara, V.; Park, D.K.; Amasia, M.; Kim, N.; Kim, J.; Kim, H.; Madou, M.; Cho, Y.K. Flow-enhanced electrochemical immunosensors on centrifugal microfluidic platforms. *Lab Chip* **2013**, *13*, 3747–3754. [CrossRef]
24. Noroozi, Z.; Kido, H.; Madou, M.J. Electrolysis-induced pneumatic pressure for control of liquids in a centrifugal system. *J. Electrochem. Soc.* **2011**, *158*, P130. [CrossRef]
25. Andreasen, S.Z.; Kwasny, D.; Amato, L.; Brøgger, A.L.; Bosco, F.G.; Andersen, K.B.; Svendsen, W.E.; Boisen, A. Integrating electrochemical detection with centrifugal microfluidics for real-time and fully automated sample testing. *RSC Adv.* **2015**, *5*, 17187–17193. [CrossRef]
26. Höfflin, J.; Delgado, S.M.T.; Sandoval, F.S.; Korvink, J.G.; Mager, D. Electrifying the disk: A modular rotating platform for wireless power and data transmission for Lab on a disk application. *Lab Chip* **2015**, *15*, 2584–2587. [CrossRef] [PubMed]
27. Martinez-Duarte, R.; Gorkin, R.A., III; Abi-Samra, K.; Madou, M.J. The integration of 3D carbon-electrode dielectrophoresis on a CD-like centrifugal microfluidic platform. *Lab Chip* **2010**, *10*, 1030–1043. [CrossRef] [PubMed]
28. Smith, S.; Mager, D.; Perebikovsky, A.; Shamloo, E.; Kinahan, D.; Mishra, R.; Torres Delgado, S.; Kido, H.; Saha, S.; Ducrée, J.; et al. CD-Based Microfluidics for Primary Care in Extreme Point-of-Care Settings. *Micromachines* **2016**, *7*, 22. [CrossRef]
29. Glynn, M.T.; Kinahan, D.J.; Ducrée, J. CD4 counting technologies for HIV therapy monitoring in resource-poor settings–state-of-the-art and emerging microtechnologies. *Lab Chip* **2013**, *13*, 2731–2748. [CrossRef]
30. Peeling, R.; Mabey, D. Point-of-care tests for diagnosing infections in the developing world. *Clin. Microbiol. Infect.* **2010**, *16*, 1062–1069. [CrossRef]
31. Land, K.J.; Boeras, D.I.; Chen, X.S.; Ramsay, A.R.; Peeling, R.W. REASSURED diagnostics to inform disease control strategies, strengthen health systems and improve patient outcomes. *Nat. Microbiol.* **2019**, *4*, 46–54. [CrossRef]
32. Bissonnette, L.; Bergeron, M.G. The genePOC platform, a rational solution for extreme point-of-care testing. *Micromachines* **2016**, *7*, 94. [CrossRef]
33. Tang, M.; Wang, G.; Kong, S.K.; Ho, H.P. A Review of Biomedical Centrifugal Microfluidic Platforms. *Micromachines* **2016**, *7*, 26. [CrossRef]
34. Strohmeier, O.; Keller, M.; Schwemmer, F.; Zehnle, S.; Mark, D.; Von Stetten, F.; Zengerle, R.; Paust, N. Centrifugal microfluidic platforms: advanced unit operations and applications. *Chem. Soc. Rev.* **2015**, *44*, 6187–6229. [CrossRef] [PubMed]
35. Ducrée, J.; Haeberle, S.; Lutz, S.; Pausch, S.; von Stetten, F.; Zengerle, R. The centrifugal microfluidic Bio-Disk platform. *J. Micromech. Microeng.* **2007**, *17*, S103. [CrossRef]
36. Gorkin, R.; Park, J.; Siegrist, J.; Amasia, M.; Lee, B.S.; Park, J.M.; Kim, J.; Kim, H.; Madou, M.; Cho, Y.K. Centrifugal microfluidics for biomedical applications. *Lab Chip* **2010**, *10*, 1758. [CrossRef] [PubMed]
37. Kong, L.X.; Perebikovsky, A.; Moebius, J.; Kulinsky, L.; Madou, M. Lab-on-a-CD: A fully integrated molecular diagnostic system. *J. Lab. Autom.* **2016**, *21*, 323–355. [CrossRef] [PubMed]
38. King, D.; O'Sullivan, M.; Ducrée, J. Optical detection strategies for centrifugal microfluidic platforms. *J. Mod. Opt.* **2014**, *61*, 85–101. [CrossRef]

39. Burger, R.; Amato, L.; Boisen, A. Detection methods for centrifugal microfluidic platforms. *Biosens. Bioelectron.* **2016**, *76*, 54–67. [CrossRef] [PubMed]
40. Hugo, S.; Land, K.; Madou, M.; Kido, H. A centrifugal microfluidic platform for point-of-care diagnostic applications. *S. Afr. J. Sci.* **2014**, *110*, 1–7. [CrossRef]
41. Nolte, D.D. Invited Review Article: Review of centrifugal microfluidic and bio-optical disks. *Rev. Sci. Instrum.* **2009**, *80*, 101101. [CrossRef]
42. Kinahan, D.J.; Kearney, S.M.; Dimov, N.; Glynn, M.T.; Ducrée, J. Event-triggered logical flow control for comprehensive process integration of multi-step assays on centrifugal microfluidic platforms. *Lab Chip* **2014**, *14*, 2249–2258. [CrossRef] [PubMed]
43. Clime, L.; Daoud, J.; Brassard, D.; Malic, L.; Geissler, M.; Veres, T. Active pumping and control of flows in centrifugal microfluidics. *Microfluid. Nanofluid.* **2019**, *23*, 29. [CrossRef]
44. Chen, J.M.; Huang, P.C.; Lin, M.G. Analysis and experiment of capillary valves for microfluidics on a rotating disk. *Microfluid. Nanofluid.* **2008**, *4*, 427–437. [CrossRef]
45. Gorkin, R.; Clime, L.; Madou, M.; Kido, H. Pneumatic pumping in centrifugal microfluidic platforms. *Microfluid. Nanofluid.* **2010**, *9*, 541–549. [CrossRef]
46. Siegrist, J.; Gorkin, R.; Clime, L.; Roy, E.; Peytavi, R.; Kido, H.; Bergeron, M.; Veres, T.; Madou, M. Serial siphon valving for centrifugal microfluidic platforms. *Microfluid. Nanofluid.* **2010**, *9*, 55–63. [CrossRef]
47. Godino, N.; Gorkin , R., III; Linares, A.V.; Burger, R.; Ducrée, J. Comprehensive integration of homogeneous bioassays via centrifugo-pneumatic cascading. *Lab Chip* **2013**, *13*, 685–694. [CrossRef]
48. Zhu, Y.; Chen, Y.; Xu, Y. Interruptible siphon valving for centrifugal microfluidic platforms. *Sens. Actuators B Chem.* **2018**, *276*, 313–321. [CrossRef]
49. Park, J.M.; Cho, Y.K.; Lee, B.S.; Lee, J.G.; Ko, C. Multifunctional microvalves control by optical illumination on nanoheaters and its application in centrifugal microfluidic devices. *Lab Chip* **2007**, *7*, 557–564. [CrossRef] [PubMed]
50. Gorkin, R.; Nwankire, C.E.; Gaughran, J.; Zhang, X.; Donohoe, G.G.; Rook, M.; O'Kennedy, R.; Ducrée, J. Centrifugo-pneumatic valving utilizing dissolvable films. *Lab Chip* **2012**, *12*, 2894–2902. [CrossRef] [PubMed]
51. Kinahan, D.J.; Early, P.L.; Vembadi, A.; MacNamara, E.; Kilcawley, N.A.; Glennon, T.; Diamond, D.; Brabazon, D.; Ducrée, J. Xurography actuated valving for centrifugal flow control. *Lab Chip* **2016**, *16*, 3454–3459. [CrossRef]
52. McArdle, H.; Jimenez-Mateos, E.M.; Raoof, R.; Carthy, E.; Boyle, D.; ElNaggar, H.; Delanty, N.; Hamer, H.; Dogan, M.; Huchtemann, T.; et al. "TORNADO"—Theranostic One-Step RNA Detector; microfluidic disc for the direct detection of microRNA-134 in plasma and cerebrospinal fluid. *Sci. Rep.* **2017**, *7*, 1750. [CrossRef]
53. Al-Faqheri, W.; Ibrahim, F.; Thio, T.H.G.; Aeinehvand, M.M.; Arof, H.; Madou, M. Development of novel passive check valves for the microfluidic CD platform. *Sens. Actuators A Phys.* **2015**, *222*, 245–254. [CrossRef]
54. Kazemzadeh, A.; Ganesan, P.; Ibrahim, F.; Aeinehvand, M.M.; Kulinsky, L.; Madou, M.J. Gating valve on spinning microfluidic platforms: A flow switch/control concept. *Sens. Actuators B Chem.* **2014**, *204*, 149–158. [CrossRef]
55. Kinahan, D.J.; Renou, M.; Kurzbuch, D.; Kilcawley, N.A.; Bailey, É.; Glynn, M.T.; McDonagh, C.; Ducrée, J. Baking powder actuated centrifugo-pneumatic valving for automation of multi-step bioassays. *Micromachines* **2016**, *7*, 175. [CrossRef] [PubMed]
56. Cai, Z.; Xiang, J.; Zhang, B.; Wang, W. A magnetically actuated valve for centrifugal microfluidic applications. *Sens. Actuators B Chem.* **2015**, *206*, 22–29. [CrossRef]
57. Cai, Z.; Xiang, J.; Wang, W. A pinch-valve for centrifugal microfluidic platforms and its application in sequential valving operation and plasma extraction. *Sens. Actuators B Chem.* **2015**, *221*, 257–264. [CrossRef]
58. Aeinehvand, M.M.; Weber, L.; Jiménez, M.; Palermo, A.; Bauer, M.; Loeffler, F.F.; Ibrahim, F.; Breitling, F.; Korvink, J.; Madou, M.; et al. Elastic reversible valves on centrifugal microfluidic platforms. *Lab Chip* **2019**, *19*, 1090–1100. [CrossRef] [PubMed]
59. Xiang, J.; Cai, Z.; Zhang, Y.; Wang, W. Wedge actuated normally-open and normally-closed valves for centrifugal microfluidic applications. *Sens. Actuators B Chem.* **2017**, *243*, 542–548. [CrossRef]
60. Aeinehvand, M.M.; Martins Fernandes, R.F.; Jiménez Moreno, M.F.; Lara Díaz, V.J.; Madou, M.; Martinez-Chapa, S.O. Aluminium valving and magneto-balloon mixing for rapid prediction of septic shock on centrifugal microfluidic platforms. *Sens. Actuators B Chem.* **2018**, *276*, 429–436. [CrossRef]

61. Kim, T.H.; Sunkara, V.; Park, J.; Kim, C.J.; Woo, H.K.; Cho, Y.K. A lab-on-a-disc with reversible and thermally stable diaphragm valves. *Lab Chip* **2016**, *16*, 3741–3749. [CrossRef]
62. Duffy, D.C.; McDonald, J.C.; Schueller, O.J.; Whitesides, G.M. Rapid prototyping of microfluidic systems in poly (dimethylsiloxane). *Anal. Chem.* **1998**, *70*, 4974–4984. [CrossRef]
63. Xia, Y.; McClelland, J.J.; Gupta, R.; Qin, D.; Zhao, X.M.; Sohn, L.L.; Celotta, R.J.; Whitesides, G.M. Replica molding using polymeric materials: A practical step toward nanomanufacturing. *Adv. Mater.* **1997**, *9*, 147–149. [CrossRef]
64. Gale, B.K.; Jafek, A.R.; Lambert, C.J.; Goenner, B.L.; Moghimifam, H.; Nze, U.C.; Kamarapu, S.K. A review of current methods in microfluidic device fabrication and future commercialization prospects. *Inventions* **2018**, *3*, 60. [CrossRef]
65. Tsao, C.W. Polymer microfluidics: Simple, low-cost fabrication process bridging academic lab research to commercialized production. *Micromachines* **2016**, *7*, 225. [CrossRef] [PubMed]
66. Faustino, V.; Catarino, S.O.; Lima, R.; Minas, G. Biomedical microfluidic devices by using low-cost fabrication techniques: A review. *J. Biomech.* **2016**, *49*, 2280–2292. [CrossRef] [PubMed]
67. Focke, M.; Stumpf, F.; Faltin, B.; Reith, P.; Bamarni, D.; Wadle, S.; Müller, C.; Reinecke, H.; Schrenzel, J.; Francois, P.; et al. Microstructuring of polymer films for sensitive genotyping by real-time PCR on a centrifugal microfluidic platform. *Lab Chip* **2010**, *10*, 2519–2526. [CrossRef]
68. Xia, Y.; Whitesides, G.M. Soft lithography. *Angew. Chem. Int. Ed.* **1998**, *37*, 550–575. [CrossRef]
69. Mcdonald, J.C.; Duffy, D.C.; Anderson, J.R.; Chiu, D.T. Review General Fabrication of microfluidic systems in poly (dimethylsiloxane). *Electrophoresis* **2000**, *21*, 27–40. [CrossRef]
70. Kricka, L.J.; Fortina, P.; Panaro, N.J.; Wilding, P.; Alonso-Amigo, G.; Becker, H. Fabrication of plastic microchips by hot embossing. *Lab Chip* **2002**, *2*, 1–4. [CrossRef]
71. Chen, Y. Applications of nanoimprint lithography/hot embossing: A review. *Appl. Phys. Mater. Sci. Process.* **2015**, *121*, 451–465. [CrossRef]
72. Attia, U.M.; Marson, S.; Alcock, J.R. Micro-injection moulding of polymer microfluidic devices. *Microfluid. Nanofluid.* **2009**, *7*, 1. [CrossRef]
73. Juang, Y.J.; Lee, L.J.; Koelling, K.W. Hot embossing in microfabrication. Part I: Experimental. *Polym. Eng. Sci.* **2002**, *42*, 539–550. [CrossRef]
74. Chou, S.Y.; Krauss, P.R. Imprint lithography with sub-10 nm feature size and high throughput. *Microelectron. Eng.* **1997**, *35*, 237–240. [CrossRef]
75. Koerner, T.; Brown, L.; Xie, R.; Oleschuk, R.D. Epoxy resins as stamps for hot embossing of microstructures and microfluidic channels. *Sens. Actuators B Chem.* **2005**, *107*, 632–639. [CrossRef]
76. Goral, V.N.; Hsieh, Y.C.; Petzold, O.N.; Faris, R.A.; Yuen, P.K. Hot embossing of plastic microfluidic devices using poly (dimethylsiloxane) molds. *J. Micromech. Microeng.* **2010**, *21*, 017002. [CrossRef]
77. Ho, C.M.B.; Ng, S.H.; Li, K.H.H.; Yoon, Y.J. 3D printed microfluidics for biological applications. *Lab Chip* **2015**, *15*, 3627–3637. [CrossRef]
78. Chen, C.; Mehl, B.T.; Munshi, A.S.; Townsend, A.D.; Spence, D.M.; Martin, R.S. 3D-printed microfluidic devices: Fabrication, advantages and limitations—A mini review. *Anal. Methods* **2016**, *8*, 6005–6012. [CrossRef]
79. Butler, J.E. Solid supports in enzyme-linked immunosorbent assay and other solid-phase immunoassays. *Methods* **2000**, *22*, 4–23. [CrossRef]
80. Bai, Y.; Koh, C.G.; Boreman, M.; Juang, Y.J.; Tang, I.C.; Lee, L.J.; Yang, S.T. Surface modification for enhancing antibody binding on polymer-based microfluidic device for enzyme-linked immunosorbent assay. *Langmuir* **2006**, *22*, 9458–9467. [CrossRef]
81. Gray, J.J. The interaction of proteins with solid surfaces. *Curr. Opin. Struct. Biol.* **2004**, *14*, 110–115. [CrossRef]
82. Lai, J.; Sunderland, B.; Xue, J.; Yan, S.; Zhao, W.; Folkard, M.; Michael, B.D.; Wang, Y. Study on hydrophilicity of polymer surfaces improved by plasma treatment. *Appl. Surf. Sci.* **2006**, *252*, 3375–3379. [CrossRef]
83. Laib, S.; MacCraith, B.D. Immobilization of biomolecules on cycloolefin polymer supports. *Anal. Chem.* **2007**, *79*, 6264–6270. [CrossRef]
84. Yuan, Y.; He, H.; Lee, L.J. Protein a-based antibody immobilization onto polymeric microdevices for enhanced sensitivity of enzyme-linked immunosorbent assay. *Biotechnol. Bioeng.* **2009**, *102*, 891–901. [CrossRef] [PubMed]

85. Miyazaki, C.; Mishra, R.; Kinahan, D.; Ferreira, M.; Ducrée, J. Polyethylene imine/graphene oxide layer-by-layer surface functionalization for significantly improved limit of detection and binding kinetics of immunoassays on acrylate surfaces. *Colloids Surf. B Biointerfaces* **2017**, *158*. [CrossRef] [PubMed]
86. Nwankire, C.E.; Venkatanarayanan, A.; Glennon, T.; Keyes, T.E.; Forster, R.J.; Ducrée, J. Label-free impedance detection of cancer cells from whole blood on an integrated centrifugal microfluidic platform. *Biosens. Bioelectron.* **2015**, *68*, 382–389. [CrossRef] [PubMed]
87. Miyazaki, C.M.; Kinahan, D.J.; Mishra, R.; Mangwanya, F.; Kilcawley, N.; Ferreira, M.; Ducrée, J. Label-free, spatially multiplexed SPR detection of immunoassays on a highly integrated centrifugal Lab-on-a-Disc platform. *Biosens. Bioelectron.* **2018**, *119*, 86–93. [CrossRef]
88. Riegger, L.; Grumann, M.; Nann, T.; Riegler, J.; Ehlert, O.; Bessler, W.; Mittenbuehler, K.; Urban, G.; Pastewka, L.; Brenner, T.; et al. Read-out concepts for multiplexed bead-based fluorescence immunoassays on centrifugal microfluidic platforms. *Sens. Actuators A Phys.* **2006**, *126*, 455–462. [CrossRef]
89. Lim, C.T.; Zhang, Y. Bead-based microfluidic immunoassays: The next generation. *Biosens. Bioelectron.* **2007**, *22*, 1197–1204. [CrossRef]
90. Li, L.; Miao, B.; Li, Z.; Sun, Z.; Peng, N. Sample-to-Answer Hepatitis B Virus DNA Detection from Whole Blood on a Centrifugal Microfluidic Platform with Double Rotation Axes. *ACS Sens.* **2019**, *4*, 2738–2745. [CrossRef]
91. Koh, C.Y.; Schaff, U.Y.; Piccini, M.E.; Stanker, L.H.; Cheng, L.W.; Ravichandran, E.; Singh, B.R.; Sommer, G.J.; Singh, A.K. Centrifugal microfluidic platform for ultrasensitive detection of botulinum toxin. *Anal. Chem.* **2015**, *87*, 922–928. [CrossRef]
92. Lee, B.S.; Lee, J.N.; Park, J.M.; Lee, J.G.; Kim, S.; Cho, Y.K.; Ko, C. A fully automated immunoassay from whole blood on a disc. *Lab Chip* **2009**, *9*, 1548–1555. [CrossRef]
93. Honda, N.; Lindberg, U.; Andersson, P.; Hoffmann, S.; Takei, H. Simultaneous Multiple Immunoassays in a Compact Disc—Shaped Microfluidic Device Based on Centrifugal Force. *Clin. Chem.* **2005**, *51*, 1955–1961. [CrossRef]
94. Van Oordt, T.; Barb, Y.; Smetana, J.; Zengerle, R.; Von Stetten, F. Miniature stick-packaging-an industrial technology for pre-storage and release of reagents in lab-on-a-chip systems. *Lab Chip* **2013**, *13*, 2888–2892. [CrossRef] [PubMed]
95. Stumpf, F.; Schwemmer, F.; Hutzenlaub, T.; Baumann, D.; Strohmeier, O.; Dingemanns, G.; Simons, G.; Sager, C.; Plobner, L.; Von Stetten, F.; et al. LabDisk with complete reagent prestorage for sample-to-answer nucleic acid based detection of respiratory pathogens verified with influenza A H3N2 virus. *Lab Chip* **2016**, *16*, 199–207. [CrossRef] [PubMed]
96. Hoffmann, J.; Mark, D.; Lutz, S.; Zengerle, R.; von Stetten, F. Pre-storage of liquid reagents in glass ampoules for DNA extraction on a fully integrated lab-on-a-chip cartridge. *Lab Chip* **2010**, *10*, 1480–1484. [CrossRef] [PubMed]
97. Kazemzadeh, A.; Eriksson, A.; Madou, M.; Russom, A. A micro-dispenser for long-term storage and controlled release of liquids. *Nat. Commun.* **2019**, *10*, 1–11. [CrossRef] [PubMed]
98. Nwankire, C.E.; Chan, D.S.S.; Gaughran, J.; Burger, R.; Gorkin, R.; Ducrée, J. Fluidic Automation of Nitrate and Nitrite Bioassays in Whole Blood by Dissolvable-Film Based Centrifugo-Pneumatic Actuation. *Sensors* **2013**, *13*, 11336–11349. [CrossRef]
99. Mishra, R.; Zapatero-Rodríguez, J.; Sharma, S.; Kelly, D.; McAuley, D.; Gilgunn, S.; O'Kennedy, R.; Ducrée, J. Automation of multi-analyte prostate cancer biomarker immunoassay panel from whole blood by minimum-instrumentation rotational flow control. *Sens. Actuators B Chem.* **2018**, *263*, 668–675. [CrossRef]
100. Kuo, J.N.; Chen, X.F. Plasma separation and preparation on centrifugal microfluidic disk for blood assays. *Microsyst. Technol.* **2015**, *21*, 2485–2494. [CrossRef]
101. Kuo, J.N.; Chen, X.F. Decanting and mixing of supernatant human blood plasma on centrifugal microfluidic platform. *Microsyst. Technol.* **2016**, *22*, 861–869. [CrossRef]
102. Lin, C.H.; Shih, C.H.; Lu, C.H. A fully integrated prothrombin time test on the microfluidic disk analyzer. *J. Nanosci. Nanotechnol.* **2013**, *13*, 2206–2212. [CrossRef]
103. Lin, C.H.; Lin, K.W.; Yen, D.; Shih, C.H.; Lu, C.H.; Wang, J.M.; Lin, C.Y. A point-of-care prothrombin time test on a microfluidic disk analyzer using alternate spinning. *J. Nanosci. Nanotechnol.* **2015**, *15*, 1401–1407. [CrossRef]

104. Park, J.M.; Kim, M.S.; Moon, H.S.; Yoo, C.E.; Park, D.; Kim, Y.J.; Han, K.Y.; Lee, J.Y.; Oh, J.H.; Kim, S.S.; et al. Fully automated circulating tumor cell isolation platform with large-volume capacity based on lab-on-a-disc. *Anal. Chem.* **2014**, *86*, 3735–3742. [CrossRef] [PubMed]
105. Kim, T.H.; Hwang, H.; Gorkin, R.; Madou, M.; Cho, Y.K. Geometry effects on blood separation rate on a rotating disc. *Sens. Actuators B Chem.* **2013**, *178*, 648–655. [CrossRef]
106. Boycott, A. Sedimentation of blood corpuscles. *Nature* **1920**, *104*, 532–532. [CrossRef]
107. Agarwal, R.; Sarkar, A.; Chakraborty, S. Interplay of Coriolis effect with rheology results in unique blood dynamics on a compact disc. *Analyst* **2019**, *144*, 3782–3789. [CrossRef] [PubMed]
108. Kinahan, D.J.; Kearney, S.M.; Glynn, M.T.; Ducrée, J. Spira mirabilis enhanced whole blood processing in a lab-on-a-disk. *Sens. Actuators A Phys.* **2014**, *215*, 71–76. [CrossRef]
109. Yu, Z.T.F.; Joseph, J.G.; Liu, S.X.; Cheung, M.K.; Haffey, P.J.; Kurabayashi, K.; Fu, J. Centrifugal microfluidics for sorting immune cells from whole blood. *Sens. Actuators B Chem.* **2017**, *245*, 1050–1061. [CrossRef] [PubMed]
110. Uddin, R.; Donolato, M.; Hwu, E.T.; Hansen, M.F.; Boisen, A. Combined detection of C-reactive protein and PBMC quantification from whole blood in an integrated lab-on-a-disc microfluidic platform. *Sens. Actuators B Chem.* **2018**, *272*, 634–642. [CrossRef]
111. Glynn, M.; Kirby, D.; Chung, D.; Kinahan, D.J.; Kijanka, G.; Ducrée, J. Centrifugo-magnetophoretic purification of CD4+ cells from whole blood toward future HIV/AIDS point-of-care applications. *J. Lab. Autom.* **2014**, *19*, 285–296. [CrossRef]
112. Moen, S.T.; Hatcher, C.L.; Singh, A.K. A Centrifugal Microfluidic Platform That Separates Whole Blood Samples into Multiple Removable Fractions Due to Several Discrete but Continuous Density Gradient Sections. *PLoS ONE* **2016**, *11*, e0153137. [CrossRef]
113. Morijiri, T.; Yamada, M.; Hikida, T.; Seki, M. Microfluidic counterflow centrifugal elutriation system for sedimentation-based cell separation. *Microfluid. Nanofluid.* **2013**, *14*, 1049–1057. [CrossRef]
114. Tefferi, A.; Hanson, C.A.; Inwards, D.J. How to interpret and pursue an abnormal complete blood cell count in adults. *Mayo Clin. Proc.* **2005**, *80*, 923–936. [CrossRef] [PubMed]
115. Park, J.M.; Lee, J.Y.; Lee, J.G.; Jeong, H.; Oh, J.M.; Kim, Y.J.; Park, D.; Kim, M.S.; Lee, H.J.; Oh, J.H.; et al. Highly efficient assay of circulating tumor cells by selective sedimentation with a density gradient medium and microfiltration from whole blood. *Anal. Chem.* **2012**, *84*, 7400–7407. [CrossRef] [PubMed]
116. Lim, M.; Park, J.; Lowe, A.C.; Jeong, H.o.; Lee, S.; Park, H.C.; Lee, K.; Kim, G.H.; Kim, M.H.; Cho, Y.K. A lab-on-a-disc platform enables serial monitoring of individual CTCs associated with tumor progression during EGFR-targeted therapy for patients with NSCLC. *Theranostics* **2020**, *10*, 5181. [CrossRef] [PubMed]
117. Kim, C.J.; Ki, D.Y.; Park, J.; Sunkara, V.; Kim, T.H.; Min, Y.; Cho, Y.K. Fully automated platelet isolation on a centrifugal microfluidic device for molecular diagnostics. *Lab Chip* **2020**, *20*, 949–957. [CrossRef] [PubMed]
118. Schaff, U.Y.; Sommer, G.J. Whole blood immunoassay based on centrifugal bead sedimentation. *Clin. Chem.* **2011**, *57*, 753–761. [CrossRef] [PubMed]
119. Mark, D.; Weber, P.; Lutz, S.; Focke, M.; Zengerle, R.; von Stetten, F. Aliquoting on the centrifugal microfluidic platform based on centrifugo-pneumatic valves. *Microfluid. Nanofluid.* **2011**, *10*, 1279–1288. [CrossRef]
120. Mark, D.; Metz, T.; Haeberle, S.; Lutz, S.; Ducrée, J.; Zengerle, R.; von Stetten, F. Centrifugo-pneumatic valve for metering of highly wetting liquids on centrifugal microfluidic platforms. *Lab Chip* **2009**, *9*, 3599–3603. [CrossRef] [PubMed]
121. Kim, T.H.; Kim, C.J.; Kim, Y.; Cho, Y.K. Centrifugal microfluidic system for a fully automated N-fold serial dilution. *Sens. Actuators B Chem.* **2018**, *256*, 310–317. [CrossRef]
122. Ducrée, J.; Brenner, T.; Haeberle, S.; Glatzel, T.; Zengerle, R. Multilamination of flows in planar networks of rotating microchannels. *Microfluid. Nanofluid.* **2006**, *2*, 78–84. [CrossRef]
123. Ducrée, J.; Haeberle, S.; Brenner, T.; Glatzel, T.; Zengerle, R. Patterning of flow and mixing in rotating radial microchannels. *Microfluid. Nanofluid.* **2006**, *2*, 97–105. [CrossRef]
124. Grumann, M.; Geipel, A.; Riegger, L.; Zengerle, R.; Ducrée, J. Batch-mode mixing on centrifugal microfluidic platforms. *Lab Chip* **2005**, *5*, 560–565. [CrossRef] [PubMed]
125. Park, S.J.; Kim, J.K.; Park, J.; Chung, S.; Chung, C.; Chang, J.K. Rapid three-dimensional passive rotation micromixer using the breakup process. *J. Micromech. Microeng.* **2003**, *14*, 6. [CrossRef]
126. Park, J.; Lee, G.H.; Park, J.Y.; Lee, J.C.; Kim, H.C. A numerical study of the Coriolis effect in centrifugal microfluidics with different channel arrangements. *Microfluid. Nanofluid.* **2016**, *20*, 65. [CrossRef]

127. Cai, Z.; Xiang, J.; Chen, H.; Wang, W. A Rapid Micromixer for Centrifugal Microfluidic Platforms. *Micromachines* **2016**, *7*, 89. [CrossRef] [PubMed]
128. Rida, A.; Lehnert, T.; Gijs, M. Microfluidic mixer using magnetic beads. In Proceedings of the 7th International Conference on Miniaturized Chemical and Biochemical Analysis Systems, Squaw Valley, California, 5–9 October 2003; pp. 579–582.
129. Burger, R.; Reith, P.; Akujobi, V.; Ducrée, J. Rotationally controlled magneto-hydrodynamic particle handling for bead-based microfluidic assays. *Microfluid. Nanofluid.* **2012**, *13*, 675–681. [CrossRef]
130. Hessel, V.; Löwe, H.; Schönfeld, F. Micromixers—A review on passive and active mixing principles. *Chem. Eng. Sci.* **2005**, *60*, 2479–2501. [CrossRef]
131. Noroozi, Z.; Kido, H.; Micic, M.; Pan, H.; Bartolome, C.; Princevac, M.; Zoval, J.; Madou, M. Reciprocating flow-based centrifugal microfluidics mixer. *Rev. Sci. Instrum.* **2009**, *80*. [CrossRef] [PubMed]
132. Noroozi, Z.; Kido, H.; Peytavi, R.; Nakajima-Sasaki, R.; Jasinskas, A.; Micic, M.; Felgner, P.L.; Madou, M.J. A multiplexed immunoassay system based upon reciprocating centrifugal microfluidics. *Rev. Sci. Instruments* **2011**, *82*. [CrossRef]
133. Hess, J.; Zehnle, S.; Juelg, P.; Hutzenlaub, T.; Zengerle, R.; Paust, N. Review on pneumatic operations in centrifugal microfluidics. *Lab Chip* **2019**, *19*, 3745–3770. [CrossRef]
134. Aeinehvand, M.M.; Ibrahim, F.; Al-Faqheri, W.; Thio, T.H.G.; Kazemzadeh, A.; Madou, M. Latex micro-balloon pumping in centrifugal microfluidic platforms. *Lab Chip* **2014**, *14*, 988–997. [CrossRef]
135. Clime, L.; Brassard, D.; Geissler, M.; Veres, T. Active pneumatic control of centrifugal microfluidic flows for lab-on-a-chip applications. *Lab Chip* **2015**, *15*, 2400–2411. [CrossRef]
136. Burger, S.; Schulz, M.; von Stetten, F.; Zengerle, R.; Paust, N. Rigorous buoyancy driven bubble mixing for centrifugal microfluidics. *Lab Chip* **2016**, *16*, 261–268. [CrossRef] [PubMed]
137. Czilwik, G.; Messinger, T.; Strohmeier, O.; Wadle, S.; Von Stetten, F.; Paust, N.; Roth, G.; Zengerle, R.; Saarinen, P.; Niittymäki, J.; et al. Rapid and fully automated bacterial pathogen detection on a centrifugal-microfluidic LabDisk using highly sensitive nested PCR with integrated sample preparation. *Lab Chip* **2015**, *15*, 3749–3759. [CrossRef] [PubMed]
138. Seo, J.H.; Park, B.H.; Oh, S.J.; Choi, G.; Kim, D.H.; Lee, E.Y.; Seo, T.S. Development of a high-throughput centrifugal loop-mediated isothermal amplification microdevice for multiplex foodborne pathogenic bacteria detection. *Sens. Actuators B Chem.* **2017**, *246*, 146–153. [CrossRef]
139. Loo, J.; Kwok, H.; Leung, C.; Wu, S.; Law, I.; Cheung, Y.; Cheung, Y.; Chin, M.; Kwan, P.; Hui, M.; et al. Sample-to-answer on molecular diagnosis of bacterial infection using integrated lab-on-a-disc. *Biosens. Bioelectron.* **2017**, *93*, 212–219. [CrossRef] [PubMed]
140. Lim, D.Y.S.; Seo, M.J.; Yoo, J.C. Optical Temperature Control Unit and Convolutional Neural Network for Colorimetric Detection of Loop-Mediated Isothermal Amplification on a Lab-On-A-Disc Platform. *Sensors* **2019**, *19*, 3207. [CrossRef] [PubMed]
141. Sayad, A.; Ibrahim, F.; Uddin, S.M.; Cho, J.; Madou, M.; Thong, K.L. A microdevice for rapid, monoplex and colorimetric detection of foodborne pathogens using a centrifugal microfluidic platform. *Biosens. Bioelectron.* **2018**, *100*, 96–104. [CrossRef]
142. Brennan, D.; Coughlan, H.; Clancy, E.; Dimov, N.; Barry, T.; Kinahan, D.; Ducrée, J.; Smith, T.J.; Galvin, P. Development of an on-disc isothermal in vitro amplification and detection of bacterial RNA. *Sens. Actuators B Chem.* **2017**, *239*, 235–242. [CrossRef]
143. Schuler, F.; Schwemmer, F.; Trotter, M.; Wadle, S.; Zengerle, R.; von Stetten, F.; Paust, N. Centrifugal step emulsification applied for absolute quantification of nucleic acids by digital droplet RPA. *Lab Chip* **2015**, *15*, 2759–2766. [CrossRef]
144. Lutz, S.; Weber, P.; Focke, M.; Faltin, B.; Hoffmann, J.; Müller, C.; Mark, D.; Roth, G.; Munday, P.; Armes, N.; et al. Microfluidic lab-on-a-foil for nucleic acid analysis based on isothermal recombinase polymerase amplification (RPA). *Lab Chip* **2010**, *10*, 887–893. [CrossRef]
145. Antunes, P.; Watterson, D.; Parmvi, M.; Burger, R.; Boisen, A.; Young, P.; Cooper, M.A.; Hansen, M.F.; Ranzoni, A.; Donolato, M. Quantification of NS1 dengue biomarker in serum via optomagnetic nanocluster detection. *Sci. Rep.* **2015**, *5*, 16145. [CrossRef] [PubMed]
146. Uddin, R.; Burger, R.; Donolato, M.; Fock, J.; Creagh, M.; Hansen, M.F.; Boisen, A. Lab-on-a-disc agglutination assay for protein detection by optomagnetic readout and optical imaging using nano- and micro-sized magnetic beads. *Biosens. Bioelectron.* **2016**, *85*, 351–357. [CrossRef]

147. Steigert, J.; Grumann, M.; Dube, M.; Streule, W.; Riegger, L.; Brenner, T.; Koltay, P.; Mittmann, K.; Zengerle, R.; Ducrée, J. Direct hemoglobin measurement on a centrifugal microfluidic platform for point-of-care diagnostics. *Sens. Actuators A Phys.* **2006**, *130–131*, 228–233. [CrossRef]
148. Grumann, M.; Steigert, J.; Riegger, L.; Moser, I.; Enderle, B.; Riebeseel, K.; Urban, G.; Zengerle, R.; Ducrée, J. Sensitivity enhancement for colorimetric glucose assays on whole blood by on-chip beam-guidance. *Biomed. Microdevices* **2006**, *8*, 209–214. [CrossRef] [PubMed]
149. Ukita, Y.; Takamura, Y. A new stroboscopic technique for the observation of microscale fluorescent objects on a spinning platform in centrifugal microfluidics. *Microfluid. Nanofluid.* **2014**, *18*, 245–252. [CrossRef]
150. Piruska, A.; Nikcevic, I.; Lee, S.H.; Ahn, C.; Heineman, W.R.; Limbach, P.A.; Seliskar, C.J. The autofluorescence of plastic materials and chips measured under laser irradiation. *Lab Chip* **2005**, *5*, 1348–1354. [CrossRef] [PubMed]
151. Hemmi, A.; Usui, T.; Moto, A.; Tobita, T.; Soh, N.; Nakano, K.; Zeng, H.; Uchiyama, K.; Imato, T.; Nakajima, H. A surface plasmon resonance sensor on a compact disk-type microfluidic device. *J. Sep. Sci.* **2011**, *34*, 2913–2919. [CrossRef]
152. Rattanarat, P.; Teengam, P.; Siangproh, W.; Ishimatsu, R.; Nakano, K.; Chailapakul, O.; Imato, T. An electrochemical compact disk-type microfluidics platform for use as an enzymatic biosensor. *Electroanalysis* **2015**, *27*, 703–712. [CrossRef]
153. Rajendran, S.T.; Scarano, E.; Bergkamp, M.H.; Capria, A.M.; Cheng, C.H.; Sanger, K.; Ferrari, G.; Nielsen, L.H.; Hwu, E.T.; Zor, K.; et al. Modular, lightweight, wireless potentiostat-on-a-disc for electrochemical detection in centrifugal microfluidics. *Anal. Chem.* **2019**, *91*, 11620–11628. [CrossRef]
154. Czugala, M.; Maher, D.; Collins, F.; Burger, R.; Hopfgartner, F.; Yang, Y.; Zhaou, J.; Ducrée, J.; Smeaton, A.; Fraser, K.J.; et al. CMAS: Fully integrated portable centrifugal microfluidic analysis system for on-site colorimetric analysis. *RSC Adv.* **2013**, *3*, 15928–15938. [CrossRef]
155. Jung, J.H.; Park, B.H.; Oh, S.J.; Choi, G.; Seo, T.S. Integrated centrifugal reverse transcriptase loop-mediated isothermal amplification microdevice for influenza A virus detection. *Biosens. Bioelectron.* **2015**, *68*, 218–224. [CrossRef] [PubMed]
156. Chang, H.C.; Chao, Y.T.; Yen, J.Y.; Yu, Y.L.; Lee, C.N.; Ho, B.C.; Liu, K.C.; Fang, J.; Lin, C.W.; Lee, J.H. A Turbidity Test Based Centrifugal Microfluidics Diagnostic System for Simultaneous Detection of HBV, HCV, and CMV. *Adv. Mater. Sci. Eng.* **2015**, *2015*. [CrossRef]
157. Liu, Q.; Zhang, X.; Chen, L.; Yao, Y.; Ke, S.; Zhao, W.; Yang, Z.; Sui, G. A sample-to-answer labdisc platform integrated novel membrane-resistance valves for detection of highly pathogenic avian influenza viruses. *Sens. Actuators B Chem.* **2018**, *270*, 371–381. [CrossRef]
158. Miao, B.; Peng, N.; Li, L.; Li, Z.; Hu, F.; Zhang, Z.; Wang, C. Centrifugal Microfluidic System for Nucleic Acid Amplification and Detection. *Sensors* **2015**, *15*, 27954–27968. [CrossRef]
159. Amasia, M.; Cozzens, M.; Madou, M.J. Centrifugal microfluidic platform for rapid PCR amplification using integrated thermoelectric heating and ice-valving. *Sens. Actuators B Chem.* **2012**, *161*, 1191–1197. [CrossRef]
160. Sayad, A.A.; Ibrahim, F.; Uddin, S.M.; Pei, K.X.; Mohktar, M.S.; Madou, M.; Thong, K.L. A microfluidic lab-on-a-disc integrated loop mediated isothermal amplification for foodborne pathogen detection. *Sens. Actuators B Chem.* **2016**, *227*, 600–609. [CrossRef]
161. Thiha, A.; Ibrahim, F. A Colorimetric Enzyme-Linked Immunosorbent Assay (ELISA) Detection Platform for a Point-of-Care Dengue Detection System on a Lab-on-Compact-Disc. *Sensors* **2015**, *15*, 11431–11441. [CrossRef] [PubMed]
162. Nwankire, C.E.; Donohoe, G.G.; Zhang, X.; Siegrist, J.; Somers, M.; Kurzbuch, D.; Monaghan, R.; Kitsara, M.; Burger, R.; Hearty, S.; et al. At-line bioprocess monitoring by immunoassay with rotationally controlled serial siphoning and integrated supercritical angle fluorescence optics. *Anal. Chim. Acta* **2013**, *781*, 54–62. [CrossRef]
163. Oh, S.J.; Park, B.H.; Jung, J.H.; Choi, G.; Lee, D.C.; Seo, T.S. Centrifugal loop-mediated isothermal amplification microdevice for rapid, multiplex and colorimetric foodborne pathogen detection. *Biosens. Bioelectron.* **2016**, *75*, 293–300. [CrossRef]
164. Mahmodi Arjmand, E.; Saadatmand, M.; Bakhtiari, M.R.; Eghbal, M. Design and fabrication of a centrifugal microfluidic disc including septum valve for measuring hemoglobin A1c in human whole blood using immunoturbidimetry method. *Talanta* **2018**, *190*, 134–139. [CrossRef]

165. Miyazaki, C.M.; Shimizu, F.M.; Ferreira, M. Surface plasmon resonance (SPR) for sensors and biosensors. In *Nanocharacterization Techniques*; Elsevier: Amsterdam, The Netherlands, 2017; pp. 183–200.
166. Yih, J.N.; Chiu, K.C.; Chou, S.Y.; Lin, C.M.; Lan, Y.S.; Chen, S.J.; Cheng, N.J. Surface plasmon resonance biosensor based on grating disc with circular fluidic channel. In Proceedings of the 2011 6th IEEE International Conference on Nano/Micro Engineered and Molecular Systems, Kaohsiung, Taiwan, 20–23 February 2011; pp. 571–574.
167. Biosurfit. Biosurfit. Available online: https://www.biosurfit.com/en/ (accessed on 26 May 2020).
168. Martin, J.; Nieuwoudt, M.; Vargas, M.; Bodley, O.; Yohendiran, T.; Oosterbeek, R.; Williams, D.; Simpson, M.C. Raman on a disc: High-quality Raman spectroscopy in an open channel on a centrifugal microfluidic disc. *Analyst* **2017**, *142*, 1682–1688. [CrossRef] [PubMed]
169. Choi, D.; Kang, T.; Cho, H.; Choi, Y.; Lee, L.P. Additional amplifications of SERS via an optofluidic CD-based platform. *Lab Chip* **2009**, *9*, 239–243. [CrossRef]
170. Morelli, L.; Serioli, L.; Centorbi, F.A.; Jendresen, C.B.; Matteucci, M.; Ilchenko, O.; Demarchi, D.; Nielsen, A.T.; Zór, K.; Boisen, A. Injection molded lab-on-a-disc platform for screening of genetically modified E. coli using liquid–liquid extraction and surface enhanced Raman scattering. *Lab Chip* **2018**, *18*, 869–877. [CrossRef] [PubMed]
171. Fan, M.; Andrade, G.F.; Brolo, A.G. A review on recent advances in the applications of surface-enhanced Raman scattering in analytical chemistry. *Anal. Chim. Acta* **2020**, *1097*, 1–29. [CrossRef] [PubMed]
172. Okamoto, S.; Ukita, Y. Automatic microfluidic enzyme-linked immunosorbent assay based on CLOCK-controlled autonomous centrifugal microfluidics. *Sens. Actuators B Chem.* **2018**, *261*, 264–270. [CrossRef]
173. Nwankire, C.E.; Czugala, M.; Burger, R.; Fraser, K.J.; O'Connell, T.M.; Glennon, T.; Onwuliri, B.E.; Nduaguibe, I.E.; Diamond, D.; Ducrée, J.; et al. A portable centrifugal analyser for liver function screening. *Biosens. Bioelectron.* **2014**, *56*, 352–358. [CrossRef] [PubMed]
174. Wang, K.; Liang, R.; Chen, H.; Lu, S.; Jia, S.; Wang, W. A microfluidic immunoassay system on a centrifugal platform. *Sens. Actuators B Chem.* **2017**, *251*, 242–249. [CrossRef]
175. Lee, W.S.; Sunkara, V.; Han, J.R.; Park, Y.S.; Cho, Y.K. Electrospun TiO 2 nanofiber integrated lab-on-a-disc for ultrasensitive protein detection from whole blood. *Lab Chip* **2015**, *15*, 478–485. [CrossRef]
176. Thio, T.H.G.; Ibrahim, F.; Al-Faqheri, W.; Soin, N.; Bador, M.K.; Madou, M. Sequential push-pull pumping mechanism for washing and evacuation of an immunoassay reaction chamber on a microfluidic CD platform. *PLoS ONE* **2015**, *10*, e0121836. [CrossRef]
177. Sanger, K.; Zor, K.; Jendresen, C.B.; Heiskanen, A.; Amato, L.; Nielsen, A.T.; Boisen, A. Lab-on-a-disc platform for screening of genetically modified E. coli cells via cell-free electrochemical detection of p-Coumaric acid. *Sens. Actuators B Chem.* **2017**, *253*, 999–1005. [CrossRef]
178. Yoon, Y.J.; Li, K.H.H.; Low, Y.Z.; Yoon, J.; Ng, S.H. Microfluidics biosensor chip with integrated screen-printed electrodes for amperometric detection of nerve agent. *Sens. Actuators B Chem.* **2014**, *198*, 233–238. [CrossRef]
179. Ronkainen, N.J.; Halsall, H.B.; Heineman, W.R. Electrochemical biosensors. *Chem. Soc. Rev.* **2010**, *39*, 1747–1763. [CrossRef] [PubMed]
180. Li, T.; Fan, Y.; Cheng, Y.; Yang, J. An electrochemical Lab-on-a-CD system for parallel whole blood analysis. *Lab Chip* **2013**, *13*, 2634–2640. [CrossRef] [PubMed]
181. Abi-Samra, K.; Kim, T.H.; Park, D.K.; Kim, N.; Kim, J.; Kim, H.; Cho, Y.K.; Madou, M. Electrochemical velocimetry on centrifugal microfluidic platforms. *Lab Chip* **2013**, *13*, 3253–3260. [CrossRef] [PubMed]
182. Abaxis, I. Piccolo Xpress. Available online: https://www.abaxis.com/medical/piccolo-xpress (accessed on 29 May 2020).
183. Bioscience, M. Revogene. Available online: https://www.meridianbioscience.com/platform/molecular/revogene (accessed on 29 May 2020).
184. DiaSorin Group, I. Simplexa. Available online: https://www.focusdx.com/integrated-cycler/dad-intl (accessed on 1 June 2020).
185. Hahn-Schickard. Hahn-Schickard. Available online: https://www.hahn-schickard.de/en/ (accessed on 27 May 2020).
186. Nguyen, T.; Duong Bang, D.; Wolff, A. 2019 novel coronavirus disease (COVID-19): paving the road for rapid detection and point-of-care diagnostics. *Micromachines* **2020**, *11*, 306. [CrossRef]
187. GmbH, S. Spindiag. Available online: https://www.spindiag.de/ (accessed on 4 June 2020).

188. Dysinger, M.; Ma, M. A Gyrolab Assay for the Quantitation of Free Complement Protein C5a in Human Plasma. *AAPS J.* **2018**, *20*, 106. [CrossRef]
189. Ghosh, J.G.; Nguyen, A.A.; Bigelow, C.E.; Poor, S.; Qiu, Y.; Rangaswamy, N.; Ornberg, R.; Jackson, B.; Mak, H.; Ezell, T.; et al. Long-acting protein drugs for the treatment of ocular diseases. *Nat. Commun.* **2017**, *8*, 14837. [CrossRef]
190. Zhu, L.; Wang, Y.; Joyce, A.; Djura, I.; Gorovits, B. Fit-for-purpose validation of a ligand binding assay for toxicokinetic study using mouse serial sampling. *Pharm. Res.* **2019**, *36*, 169. [CrossRef]
191. Lee, B.S.; Lee, Y.U.; Kim, H.S.; Kim, T.H.; Park, J.; Lee, J.G.; Kim, J.; Kim, H.; Lee, W.G.; Cho, Y.K. Fully integrated lab-on-a-disc for simultaneous analysis of biochemistry and immunoassay from whole blood. *Lab Chip* **2011**, *11*, 70–78. [CrossRef]
192. Liao, T.; Wang, X.; Donolato, M.; Harris, E.; Cruz, M.M.; Balmaseda, A.; Wang, R.Y. Evaluation of ViroTrack Sero Zika IgG/IgM, a New Rapid and Quantitative Zika Serological Diagnostic Assay. *Diagnostics* **2020**, *10*, 372. [CrossRef]
193. Alejo-Cancho, I.; Navero-Castillejos, J.; Peiró-Mestres, A.; Albarracín, R.; Barrachina, J.; Navarro, A.; Gonzalo, V.; Pastor, V.; Muñoz, J.; Martínez, M.J. Evaluation of a novel microfluidic immuno-magnetic agglutination assay method for detection of dengue virus NS1 antigen. *PLoS Negl. Trop. Dis.* **2020**, *14*, e0008082. [CrossRef]
194. Keller, M.; Wadle, S.; Paust, N.; Dreesen, L.; Nuese, C.; Strohmeier, O.; Zengerle, R.; Von Stetten, F. Centrifugo-thermopneumatic fluid control for valving and aliquoting applied to multiplex real-time PCR on off-the-shelf centrifugal thermocycler. *RSC Adv.* **2015**, *5*, 89603–89611. [CrossRef]
195. Strohmeier, O.; Laßmann, S.; Riedel, B.; Mark, D.; Roth, G.; Werner, M.; Zengerle, R.; von Stetten, F. Multiplex genotyping of KRAS point mutations in tumor cell DNA by allele-specific real-time PCR on a centrifugal microfluidic disk segment. *Microchim. Acta* **2014**, *181*, 1681–1688. [CrossRef]
196. Choi, G.; Prince, T.; Miao, J.; Cui, L.; Guan, W. Sample-to-answer palm-sized nucleic acid testing device towards low-cost malaria mass screening. *Biosens. Bioelectron.* **2018**, *115*, 83–90. [CrossRef] [PubMed]
197. Huang, G.; Huang, Q.; Xie, L.; Xiang, G.; Wang, L.; Xu, H.; Ma, L.; Luo, X.; Xin, J.; Zhou, X.; et al. A rapid, low-cost, and microfluidic chip-based system for parallel identification of multiple pathogens related to clinical pneumonia. *Sci. Rep.* **2017**, *7*, 6441. [CrossRef] [PubMed]
198. Lee, A.; Park, J.; Lim, M.; Sunkara, V.; Kim, S.Y.; Kim, G.H.; Kim, M.H.; Cho, Y.K. All-in-one centrifugal microfluidic device for size-selective circulating tumor cell isolation with high purity. *Anal. Chem.* **2014**, *86*, 11349–11356. [CrossRef] [PubMed]
199. Kim, T.H.; Lim, M.; Park, J.; Oh, J.M.; Kim, H.; Jeong, H.; Lee, S.J.; Park, H.C.; Jung, S.; Kim, B.C.; et al. FAST: Size-selective, clog-free isolation of rare cancer cells from whole blood at a liquid–liquid interface. *Anal. Chem.* **2017**, *89*, 1155–1162. [CrossRef] [PubMed]
200. Clinomics Inc. Clinomics. Available online: http://clinomics.com/en/main (accessed on 28 May 2020).
201. Kim, H.; Lim, M.; Kim, J.Y.; Shin, S.J.; Cho, Y.K.; Cho, C.H. Circulating Tumor Cells Enumerated by a Centrifugal Microfluidic Device as a Predictive Marker for Monitoring Ovarian Cancer Treatment: A Pilot Study. *Diagnostics* **2020**, *10*, 249. [CrossRef]
202. Torres Delgado, S.M.; Kinahan, D.J.; Nirupa Julius, L.A.; Mallette, A.; Ardila, D.S.; Mishra, R.; Miyazaki, C.M.; Korvink, J.G.; Ducrée, J.; Mager, D. Wirelessly powered and remotely controlled valve-array for highly multiplexed analytical assay automation on a centrifugal microfluidic platform. *Biosens. Bioelectron.* **2018**, *109*, 214–223. [CrossRef]
203. Delgado, S.M.T.; Kinahan, D.J.; Sandoval, F.S.; Julius, L.A.N.; Kilcawley, N.A.; Ducrée, J.; Mager, D. Fully automated chemiluminescence detection using an electrified-Lab-on-a-Disc (eLoaD) platform. *Lab Chip* **2016**, *16*, 4002–4011. [CrossRef]
204. Delgado, S.M.T.; Korvink, J.G.; Mager, D. The eLoaD platform endows centrifugal microfluidics with on-disc power and communication. *Biosens. Bioelectron.* **2018**, *117*, 464–473. [CrossRef] [PubMed]
205. Bauer, M.; Bartoli, J.; Martinez-Chapa, S.O.; Madou, M. Wireless electrochemical detection on a microfluidic compact disc (CD) and evaluation of redox-amplification during flow. *Micromachines* **2019**, *10*, 31. [CrossRef] [PubMed]
206. Tavakoli, H.; Zhou, W.; Ma, L.; Perez, S.; Ibarra, A.; Xu, F.; Zhan, S.; Li, X. Recent advances in microfluidic platforms for single-cell analysis in cancer biology, diagnosis and therapy. *TrAC Trends Anal. Chem.* **2019**, *117*, 13–26. [CrossRef] [PubMed]

207. Dixon, C.; Ng, A.H.; Fobel, R.; Miltenburg, M.B.; Wheeler, A.R. An inkjet printed, roll-coated digital microfluidic device for inexpensive, miniaturized diagnostic assays. *Lab Chip* **2016**, *16*, 4560–4568. [CrossRef]
208. Hin, S.; Paust, N.; Keller, M.; Rombach, M.; Strohmeier, O.; Zengerle, R.; Mitsakakis, K. Temperature change rate actuated bubble mixing for homogeneous rehydration of dry pre-stored reagents in centrifugal microfluidics. *Lab Chip* **2018**, *18*, 362–370. [CrossRef]
209. Costantini, M.; Jaroszewicz, J.; Kozoń, Ł.; Szlązak, K.; Święszkowski, W.; Garstecki, P.; Stubenrauch, C.; Barbetta, A.; Guzowski, J. 3D-Printing of Functionally Graded Porous Materials Using On-Demand Reconfigurable Microfluidics. *Angew. Chem. Int. Ed.* **2019**, *58*, 7620–7625. [CrossRef]
210. Arango, Y.; Temiz, Y.; Gökçe, O.; Delamarche, E. Electrogates for stop-and-go control of liquid flow in microfluidics. *Appl. Phys. Lett.* **2018**, *112*, 153701. [CrossRef]
211. Alazzam, A.; Alamoodi, N. Microfluidic Devices with Patterned Wettability Using Graphene Oxide for Continuous Liquid–Liquid Two-Phase Separation. *ACS Appl. Nano Mater.* **2020**, *3*, 3471–3477. [CrossRef]
212. Wang, Y.; Toyoda, K.; Uesugi, K.; Morishima, K. A simple micro check valve using a photo-patterned hydrogel valve core. *Sens. Actuators A Phys.* **2020**, *304*, 111878. [CrossRef]
213. Van Nguyen, H.; Nguyen, V.D.; Nguyen, H.Q.; Chau, T.H.T.; Lee, E.Y.; Seo, T.S. Nucleic acid diagnostics on the total integrated lab-on-a-disc for point-of-care testing. *Biosens. Bioelectron.* **2019**, *141*, 111466. [CrossRef]
214. Brassard, D.; Geissler, M.; Descarreaux, M.; Tremblay, D.; Daoud, J.; Clime, L.; Mounier, M.; Charlebois, D.; Veres, T. Extraction of nucleic acids from blood: unveiling the potential of active pneumatic pumping in centrifugal microfluidics for integration and automation of sample preparation processes. *Lab Chip* **2019**, *19*, 1941–1952. [CrossRef] [PubMed]
215. Moeller, M.E.; Fock, J.; Pah, P.; Veras, A.D.L.C.; Bade, M.; Donolato, M.; Israelsen, S.B.; Eugen-Olsen, J.; Benfield, T.; Engsig, F.N. Evaluation of commercially available immuno-magnetic agglutination and enzyme-linked immunosorbent assays for rapid point-of-care diagnostics of COVID-19. *medRxiv* **2020**. [CrossRef]
216. Vasilescu, S.A.; Bazaz, S.R.; Jin, D.; Shimoni, O.; Warkiani, M.E. 3D printing enables the rapid prototyping of modular microfluidic devices for particle conjugation. *Appl. Mater. Today* **2020**, *20*, 100726. [CrossRef]
217. Mitxelena-Iribarren, O.; Zabalo, J.; Arana, S.; Mujika, M. Improved microfluidic platform for simultaneous multiple drug screening towards personalized treatment. *Biosens. Bioelectron.* **2019**, *123*, 237–243. [CrossRef]
218. Zhang, Y.; Xiang, J.; Wang, Y.; Qiao, Z.; Wang, W. A 3D printed centrifugal microfluidic platform for spilled oil enrichment and detection based on solid phase extraction (SPE). *Sens. Actuators B Chem.* **2019**, *296*, 126603. [CrossRef]
219. Weisgrab, G.; Ovsianikov, A.; Costa, P.F. Functional 3D printing for microfluidic chips. *Adv. Mater. Technol.* **2019**, *4*, 1900275. [CrossRef]
220. Bazaz, S.R.; Rouhi, O.; Raoufi, M.A.; Ejeian, F.; Asadnia, M.; Jin, D.; Warkiani, M.E. 3D Printing of Inertial Microfluidic Devices. *Sci. Rep.* **2020**, *10*, 1–14.
221. Carrell, C.; Kava, A.; Nguyen, M.; Menger, R.; Munshi, Z.; Call, Z.; Nussbaum, M.; Henry, C. Beyond the lateral flow assay: A review of paper-based microfluidics. *Microelectron. Eng.* **2019**, *206*, 45–54. [CrossRef]

 processes

Article

Controlling Nanoparticle Formulation: A Low-Budget Prototype for the Automation of a Microfluidic Platform

Dominik M. Loy [1,*], Rafał Krzysztoń [2,3], Ulrich Lächelt [1], Joachim O. Rädler [2,3] and Ernst Wagner [1]

1 Department of Pharmacy, Ludwig-Maximilians-Universität München, Butenandtstr. 5-13, 81377 Munich, Bavaria, Germany; ulrich.laechelt@cup.uni-muenchen.de (U.L.); ernst.wagner@cup.uni-muenchen.de (E.W.)
2 Faculty of Physics, Ludwig-Maximilians-Universität München, Geschwister-Scholl-Platz 1, 80539 Munich, Bavaria, Germany; rafal.krzyszton@stonybrook.edu (R.K.); raedler@lmu.de (J.O.R.)
3 Graduate School of Quantitative Biosciences (QBM), Ludwig-Maximilians-Universität München, Geschwister-Scholl-Platz 1, 80539 Munich, Bavaria, Germany
* Correspondence: dominik.loy@cup.uni-muenchen.de

Abstract: Active pharmaceutical ingredients (API) with suboptimal pharmacokinetic properties may require formulation into nanoparticles. In addition to the quality of the excipients, production parameters are crucial for producing nanoparticles which reliably deliver APIs to their target. Microfluidic platforms promise increased control over the formulation process due to the decreased degrees of freedom at the micro- and nanoscale. Publications about these platforms usually provide only limited information about the soft- and hardware required to integrate the microfluidic chip seamlessly into an experimental set-up. We describe a modular, low-budget prototype for microfluidic mixing in detail. The prototype consists of four modules. The control module is a raspberry pi executing customizable python scripts to control the syringe pumps and the fraction collector. The feeding module consists of up to three commercially available, programable syringe pumps. The formulation module can be any macro- or microfluidic chip connectable to syringe pumps. The collection module is a custom-built fraction collector. We describe each feature of the working prototype and demonstrate its power with polyplexes formulated from siRNA and two different oligomers that are fed to the chip at two different stages during the assembly of the nanoparticles.

Keywords: nanoparticle; lipoplex; polyplex; raspberry pi; siRNA; python; microfluidics

Citation: Loy, D.M.; Krzysztoń, R.; Lächelt, U.; Rädler, J.O.; Wagner, E. Controlling Nanoparticle Formulation: A Low-Budget Prototype for the Automation of a Microfluidic Platform. *Processes* **2021**, *9*, 129. https://doi.org/10.3390/pr9010129

Received: 29 November 2020
Accepted: 5 January 2021
Published: 8 January 2021

1. Introduction

Packaging active pharmaceutical ingredients (API) into nanoparticles can alter the pharmacokinetic properties of drugs fundamentally [1–3]. It is a well-established strategy to improve the biodistribution of small molecule drugs like paclitaxel [4] as well as larger, oligomeric drugs like nucleic acids [5,6].

Many methods have been developed to produce nanoparticles containing the target API either with the top-down [7–11] or the bottom-up approach [12]. Mixing cationic oligo- or polymeric excipients with negatively charged nucleic acids to induce nanoparticle formation, for example, is an established method from the bottom-up approach [13]. These nanoparticles are named polyplexes [14].

The properties of polyplexes and ultimately their effectiveness is determined by their individual components and the formulation process parameters [15].

On the one hand, polyplex properties can be controlled by the application of solid-phase supported synthesis (SPSS) [16] to precisely design the chemical components of the polyplex. This technique enables the synthesis of oligomers with defined structures, for example, sequence defined oligo(ethanamino)amides [17]. These oligoamides form the backbone of more elaborate structures with additional functional elements integrated at exactly defined places in the oligomer sequence [18,19].

On the other hand, the formulation process as well exerts a decisive influence on the polyplex characteristics. The polyelectrolyte complex formation is thermodynamically favorable due to a high entropy gain from counter ion released during complexation [20]. Nevertheless, formulation parameters determine the polyplex properties due to kinetic control over the complex formation process [21].

These considerations indicate that both the production process and the defined polyplex components are important for the therapeutic success of the final formulation. One approach to reduce unintended variation of process parameters is the automation of the process. For example, particle properties can be improved measurably by controlling the educt feeding rates to a T-junction to produce lipoplexes [22] or polyplexes [23].

Additionally, transferring the mixing process to the micrometer scale reduces the degrees of freedom of the system [24]. At this scale, forces from interfaces greatly surpass inertial forces that dominate the macro scale [25]. These effects are exploited in microfluidic devices and potentially lead to an increased control over the mixing process itself. Many studies have demonstrated that methods based on microfluidics can improve the physicochemical properties of nanoparticles [26–29]. In our previous publication [30], we demonstrated that the application of a microfluidic chip increased control over the sequential polyplex formulation process by exploiting the advantages of solvent exchange in combination with flow focusing inside the microchannel. The educts and the production method are depicted in Figure 1. Sequence-defined formulation components were a lipo-oligomer for nanoparticle core formation [31], a lipid anchor-polyethylene glycol (PEG)-ligand for coating [30], and siRNA. Schematics of the utilized microfluidic chips (Supplemental Figures S1 and S2) are presented in the Supplemental Information. The resulting multi-component polyplexes were well-defined due to the increased level of control over the formulation process and allowed the establishment of structure-function relationships between the PEG-ligand length and the siRNA transfection efficiency [30].

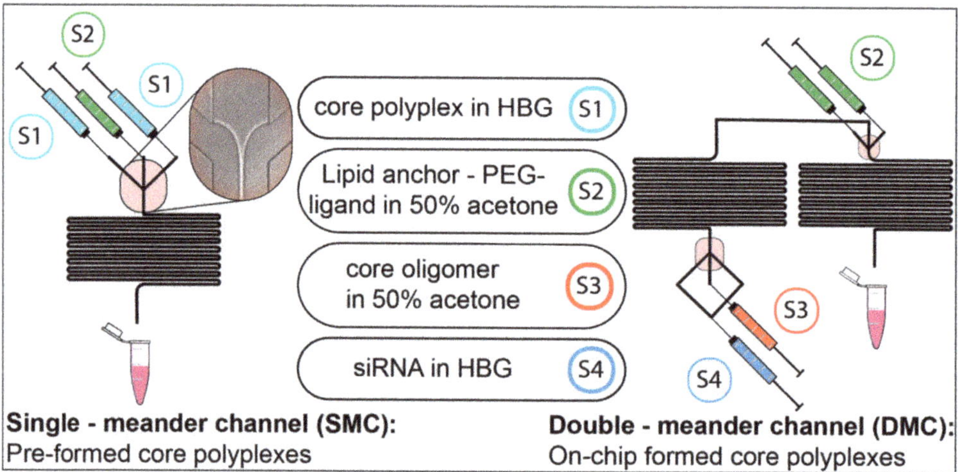

Figure 1. Production methods for polyplexes from oligomers and siRNA. Formulation components used are depicted with the id label of their corresponding syringe (S1–S4). Two different channels were used to produce nanoparticles during solvent exchange, a single meander channel (left) and a double meander channel (right). In the single meander channel, pre-assembled core particles (S1) were mixed with lipid anchor polyethylene glycol (PEG)-ligand oligomer (S2). In the double meander channel, the polyplex was assembled by microfluidics in two subsequent steps from its starting components (S3, S4, S2). PEG: polyethylene glycol. HBG: HEPES buffered glucose pH 7.4. Core oligomer (S3): cationic lipo-oligomer. Lipid anchor-PEG-ligand (S2): oligomer containing fatty acids to hydrophobically adsorb to the core oligomer, a PEG chain and folic acid as ligand for the folate receptor. Core polyplex (S1): polyplex from siRNA (S4) + core oligomer (S3). Reproduced with modification from Loy et al., 2019 [30], https://doi.org/10.7717/peerj-matsci.1/fig-1. under copyright permission from PEERJ.

Exploiting the benefits of automated nanoparticle production systems is often associated with significant investments in hardware, software, and development work to integrate any microfluidic platform into a typical lab environment. To ameliorate this problem, we have developed a modular, low budget system around the microfluidic chips from our previous publication [30]. The system can easily integrate most microfluidic platforms driven by syringe pumps into its setup. Here, we describe the set-up of the system and its application in detail using educts from our previous publication as an example.

2. Results

The complete nanoparticle production system is depicted in Figure 2. It consists of four modules that can be used independently: the feeding module—up to three programmable syringe pumps—is responsible for supplying educts to the formulation module, which can be any macro or microfluidic chip. The collection module—a custom-built fraction collector—is responsible for collecting the final product into standardized well plates. The control module is a remotely accessible raspberry pi which controls the syringe pumps via a Recommended Standard 232 (RS232) interface and the fraction collector via the general-purpose input/output (GPIO) pins. The design of the fraction collector as well as the python program code are published together with this paper on GitHub [32]. This setup allows the employment of most microfluidic chips while additionally providing the ability to sample the product from the chip directly into standardized well plates. We describe all modules in detail in the following sections.

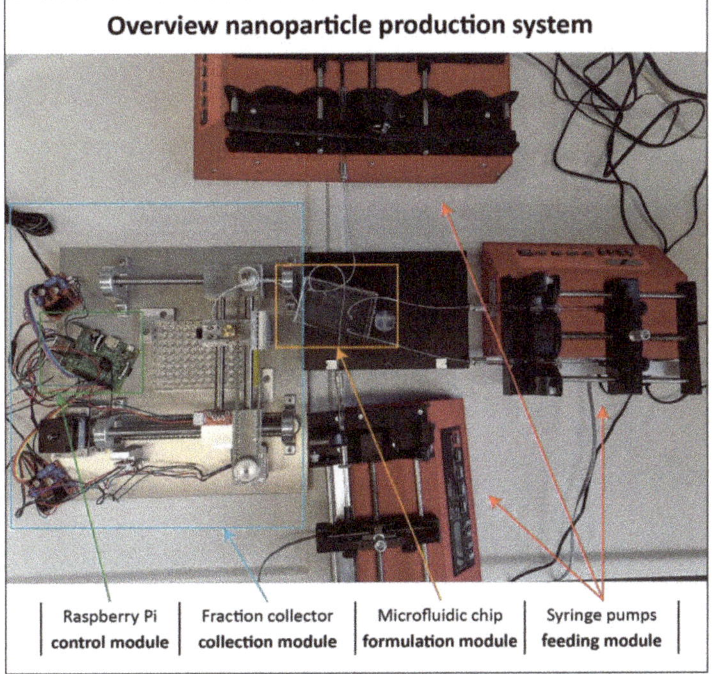

Figure 2. Overview over the nanoparticle production system. The system consists of four modules that can be used independently. The control module (green) is a raspberry pi which controls the collection module (blue), a custom-built fraction collector, via its GPIO pins. The raspberry pi controls the feeding module (red) via the RS232 interface. It is assembled from up to three syringe pumps. The formulation module can be any macro or microfluidic chip.

2.1. Control Module

The control module is a raspberry pi model 3B. It is a small and inexpensive single-board computer with open hardware, e.g., general-purpose input/output (GPIO) pins. We used the open source operating system (OS) Raspbian version 9 (Stretch) with the open source programming language Python version 3.7.3 for this system. The python code for controlling the feeding module and the collection module is available on our GitHub repository [32].

We chose this computer because of several features. First, its size together with the ability to access it remotely over the network increases the mobility of the complete device. Second, its various interfaces enable the communication with the feeding module via a USB to RS232 interface as well as the communication of the collection module via the GPIO pins. Third, it can be easily replaced by any other computer from the raspberry pi family since it is cheap and the OS together with all data, e.g., the logs of each experiment, are stored on a micro SD card. Therefore, changing the control module, e.g., because it was damaged or more computing power is needed, is only a matter of switching the SD card into the new device.

2.2. Feeding Module

The feeding module consists of up to three syringe pumps that are daisy-chained to the raspberry pi via a R232 to USB interface. Here, we used LA120, LA122, and LA160 from Landgraf Laborsysteme HLL GmbH. LA120 and LA160 are standard syringe pumps with two and six channels, respectively. LA122 is microfluidic syringe pump with two channels which is especially suited for dispensing smaller volumes due to its higher precision.

In principle, any syringe pump can be integrated into the system if it satisfies the following prerequisites: First, the pumps must have an interface that can be connected to the control module, e.g., the RS232 serial interface. Second, the pump must be programmable. In order to reduce the risk of interferences during particle production, the complete program is written to the pumps in advance and the pumps execute the production program independently. If a pump with a different command structure is integrated into the system, however, commands sent to the pump must be adjusted. A detailed description on changing commands sent to the pumps can be found in the Supplemental Information (3.2.3. Module: Module_pumps.py).

The control program of the feeding modules consists of six modules that are described in detail in the Supplemental Information (3.2. Description of the python modules) together with a Unified Modeling Language (UML) class diagram to illustrate the dependencies between the classes of the modules (Figure S3). The main module that calls the required functions from the respective modules to execute a certain pumping program is called 'main.py.' We provide a library with different 'main[...].py' modules. If the module 'main.py' is executed, the user will be asked to input all parameters during runtime, e.g., flow rates and volumes. If one of the modules 'main_[...]_automated.py' is executed, the parameters defined in the module will be used to run the pre-defined pumping program without requiring any user input. These modules serve as examples of how to define target variables and how to customize the main module. The code of the 'main.py' module is described in the Supplemental Information (3.2.7. Modules main[...].py).

Several features are implemented in the control program to simplify the employment of different formulation modules, to document experiments, and to save educts: first, formulation module specifications are loaded into the program during runtime from a simple text file. In order to employ a new channel, an updated text file needs to be supplied to the program. A detailed description on adding new formulation module specifications can be found in the Supplemental Information (3.2.1 Module: channels.py). Second, a logging function was integrated into the program, which writes every event and its timestamp to a text file stored on the control module. This log can be used for documenting and for troubleshooting purposes. Third, the implementation of ramping and purging capabilities reduces the waste of educts to a minimum. When large flow rate changes occur (e.g., when a pump is started), the system needs some time to adapt to the increased pressure. This can lead to the retardation of educts due to the elasticity of the system. Bringing educts efficiently (i.e., without wasting time or educts) to the mixing zone without involuntarily changing the volume ratios is challenging especially at the beginning of a new run. The easiest solution would be to use the flow rates of the first experiment to pump all educts to the mixing zone. Applying this strategy, however, increases waste of time and educts in relation to the flow rate differences between the educts. Ramping all educts to the mixing zone without changing the mean flow rate alleviates this problem and prevents unnecessary waste.

Additionally, employing the ramping protocol can reduce the backflow from the syringe pumps. During transition from preparations to formulation, the ramping protocol ensures a smooth transition between flow rate changes and keeps the overall flow rate constant, minimizing pressured changes that can provoke backflows. Moreover, when flow rates need to be changed during the formulation of the product, the program automatically inserts an overlap volume between those two fractions to allow some time for the flow to stabilize again. The overlap volume can be adjusted according to the magnitude of the flow rate changes. If large flow rate changes take place (e.g., when flow rates between slow and fast pumping pumps are interchanged), the modularity of the program allows another execution of the ramping protocol. Furthermore, fractions affected by backpressure instabilities can automatically be excluded using the collection module.

The ramping program is described in the Supplemental Information (3.2.5. Module: ramping_class.py).

The purging functions enable the user to choose the least expensive reagent to purge the product from the channel after the experiment. A detailed description of this function can be found in the Supplemental Information (3.2.6. Module: mixing_class.py).

A flowchart describing the workflow from starting the system to collecting the final product(s) and resetting the system to its original state is shown in the Supplemental Information (Figure S4).

2.3. Formulation Module

The formulation module can be any micro- or macrofluidic chip that is connectable to syringe pumps. In our prototype, we employed two different microfluidic chips that are based on the design from Krzysztoń et al. [29]. These chips are made from polydimethylsiloxane (PDMS) bonded to glass slides. Both chips exploit the advantages of solvent exchange in combination with flow-focusing inside the microchannel to produce polyplexes from siRNA and polycationic oligomers. The layout of both chips together with the utilized educts is shown in Figure 1. The single meander channel (SMC) employs the design of a Y-junction followed by a long meandering channel section while the double meander channel (DMC) features two successive Y-junctions followed by their respective meandering section which allows the assembly of polyplexes in two consecutive steps. Detailed schematics of both channels are shown in the Supplemental Information (SMC: Figure S1, DMC: Figure S2). The chips were made from polydimethylsiloxane (PDMS) bonded to glass slides. Wang et al. [33] have made suggestions to increase durability of these chips. We used both chips in our previous publication to produce well-defined, multicomponent polyplexes that allowed the establishment of structure-function relationships between PEG-ligand length and transfection efficiency due to the increased level of control over the formulation process [30]. Here, we only show data produced with the DMC to highlight the potential of the device to produce sophisticated formulations. For a comparison of the core formulation (siRNA and CO) prepared by the SMC, at a T-junction, or by rapid pipetting, please see Figure 2 in our previous publication [30].

2.4. Collection Module

The design is based on previously published work [34]. It was optimized for greater robustness and user safety, especially by choosing aluminum to decrease wear, increase resistance to common solvents (except acids), and to increase the accuracy of fit of the machine. Increased user safety was realized by including stop switches into the design.

The fraction collector is controlled by a raspberry pi 3, model B running Raspbian GNU/Linux 9 (stretch). The raspberry pi controls the fraction collector via input/output (GPIO) pins. Figure 3 shows an overview of the complete fraction collector (Figure 3A), the wiring of each component (Figure 3B,D), and the GPIO pin assignment (Figure 3C).

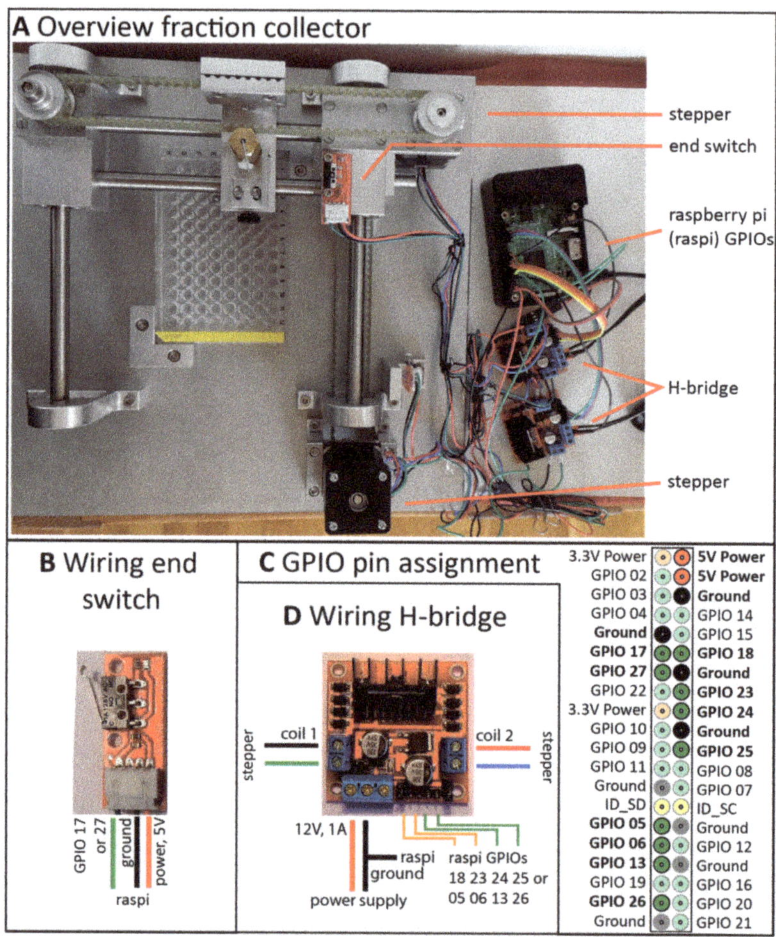

Figure 3. Overview over the fraction collector. (**A**) Overview fraction collector. (**B**) Wiring of the end switches. The end switches are supplied with 5V power from the pins of the raspberry pi, and the signal is sent from the switches to GPIO 17 or 27 (green wire). (**C**) GPIO pin assignment. Schematics of the GPIO pins of the raspberry pi. Saturated colors and bold script indicate utilized pins. (**D**) Wiring of the H-bridge. Each H-bridge controls one stepper motor. Power is supplied by a 12 V, 1 A switching power supply and routed to each stepper motor by four output wires (coil one: black and green wires; coil two: red and blue wires). Power distribution is controlled by the GPIO pins. GPIOs 18 and 23 or 05 and 06 (orange wires) control the direction of coil one, while GPIOs 24 and 25 or 13 and 26 (light green wires) control the direction of coil two.

The control program for fraction collector is an independent piece of software. This approach allows the integration of the collector control software into the pumping program but enables usage of this device with other, non-automated processes, as well. It consists of three modules which are described in detail in the Supplemental Information (4.2. Description of python modules). The main module is called 'main.py.' It calls the required functions from the respective modules to execute a certain collection program. The module serves as example how to define target variables and how to customize the collection program. A video documenting the execution of the 'main.py' module can be found on GitHub [32]. The code of the 'main.py' module is described in the Supplemental Information (4.2.3. Module: main.py). A complete list of all classes and functions of the modules can be found on GitHub, as well [32]. The dependencies between the classes of the program are depicted in a UML class diagram in the Supplemental Information (Figure S3).

2.5. Application for Polyplex Formation

In the following, we highlight the importance of precisely defined process parameters and show the formulation of three component polyplexes with the automated nanoparticle production system.

Figure 4 demonstrates the influence of formulation parameters on polyplexes formulated from two components by rapid pipetting. By adjusting the volume ratios and the mixing order of the oligomer and siRNA solutions, significant changes in size (hydrodynamic diameter) and polydispersity index (PDI) can be achieved.

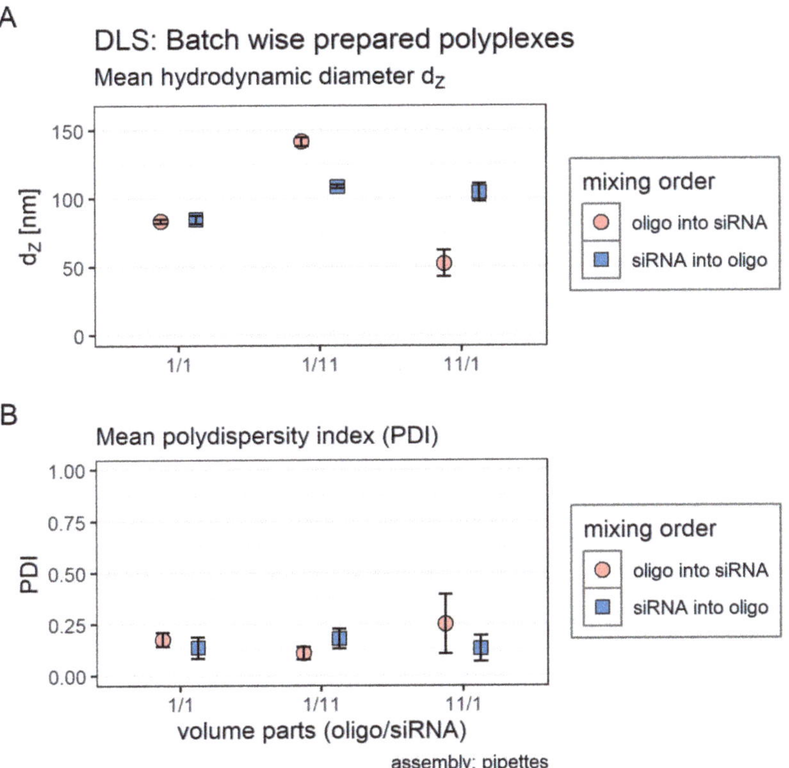

Figure 4. Influence of formulation conditions on manually prepared polyplexes (oligomer CO + siRNA). Dynamic light scattering (DLS) data are represented as the mean of three measurements. Color and shape encode the mixing order. The volume parts of both educts are denoted on the x-axis. Total volume was 70 µL for each solution. That means, for example, that 64.2 µL of a diluted oligomer solution (oligo) was pipetted to 5.8 µL of a concentrated siRNA solution (light red dot, volume parts 11/1). Blue square: siRNA was pipetted into an oligomer solution. Red dot: The oligomer solution was pipetted into a siRNA solution. (**A**) Mean hydrodynamic diameter (z-average). (**B**) Mean polydispersity index (PDI). Statistics: Error bars correspond to 95% confidence intervals. N = 3.

For this experiment we produced polyplexes using siRNA and a core oligomer (CO) [30,31]. Light red dots present data obtained after the oligomer solution was pipetted into the siRNA solution, and blue squares represent the result after pipetting the siRNA into the oligomer solution. The number written on the x-axis denotes the volume parts of the two solutions in the final solution. The final volume of each solution was always 70 µL. For example, data resulting in the third light red dot (11/1 on the x-axis) were obtained from 64.2 µL oligomer solution that was pipetted to 5.8 µL siRNA solution.

Mixtures of equal volumes of educts solutions produced comparable hydrodynamic diameters (Figure 4A, circle: 82.5 ± 2.7 nm, square: 84.4 ± 3.1 nm) and PDIs (Figure 4B, circle: 0.151 ± 0.058, square: 0.136 ± 0.052) independent of the mixing order. If unequal volumes were mixed, however, the mixing order influenced particle characteristics significantly. Pipetting a smaller volume of the oligomer solution into a larger volume of the siRNA solution (1/11 on the x-axis) produced larger polyplexes (141.0 ± 3.2 nm) with a smaller PDI (0.109 ± 0.030), while mixing siRNA solution to an oligomer solution produced smaller particles (108.0 ± 1.4 nm) with a larger PDI (0.180 ± 0.048).

Pipetting a larger volume of the oligomer solution to a smaller volume of the siRNA solution (11/1 on the x-axis) produced very small particles (Figure 4A, 52.5 ± 9.7 nm) with a larger PDI (Figure 4B, 0.249 ± 0.144). Mixing diluted siRNA solution to concentrated oligomer solution produced slightly larger particles (104.0 ± 6.5 nm) with a comparable PDI (0.131 ± 0.063) in comparison to polyplexes from mixtures of equal volumes. The 95% confidence intervals from the z-average as well as from the PDI, however, were very large, indicating the presence of particles from different size classes.

Nanoparticles formulated from more than two components usually require increased control over the production process. Figure 5 highlights this critical issue. The formulation (siRNA/CO; 1/1 on the x-axis, light red dot) described in Figure 4 was further modified with a third oligomer that contributes shielding and targeting features to the nanoparticle. It consists of a lipid anchor for integrating into the core particle, a PEG12 chain for shielding purposes and an azide moiety that allows the simple addition of further shielding and targeting ligands via strain-promoted azide-alkyne click chemistry [35]. Here, we utilize the two simplest versions of the lipid anchor oligomer with a free azide moiety and with (LPOE) or without (LPO) two additional glutamic acids (E). The sequences of all oligomers are depicted in the Supplemental Information (Figure S7). Results from in vitro experiments with these three component polyplexes with PEG-Folic acid ligands with 12 to 60 ethylene oxide repetitions can be found in our previous publication [30]. In the same publication, we compared the influence of the production method on the biological activity of two component polyplexes. We were able to demonstrate comparable biological activity in vitro [30].

In Figure 5A,B, equal volumes of the three educts were mixed sequentially by rapid pipetting according to the order of appearance denoted on the x-axis. Color and shape indicate if LPO or LPOE was used. Manual production of three component polyplexes from CO, siRNA, and LPO yielded polyplexes with suboptimal hydrodynamic diameters and PDIs regardless of mixing order (CO + siRNA + LPO: d_Z = 416.2 ± 62.5 nm, PDI = 0.711 ± 0.233; CO + LPO + siRNA: d_Z = 466.7 ± 33.1 nm, PDI = 0.792 ± 0.068). When LPO was replaced with LPOE, the mean hydrodynamic diameter of the polyplexes was reduced to acceptable levels, but the mean PDI was still too large (CO + siRNA + LPOE: d_Z = 128.2 ± 22.4 nm, PDI = 0.468 ± 0.101; CO + LPOE + siRNA: d_Z = 114.5 ± 1.5 nm, PDI = 0.428 ± 0.058).

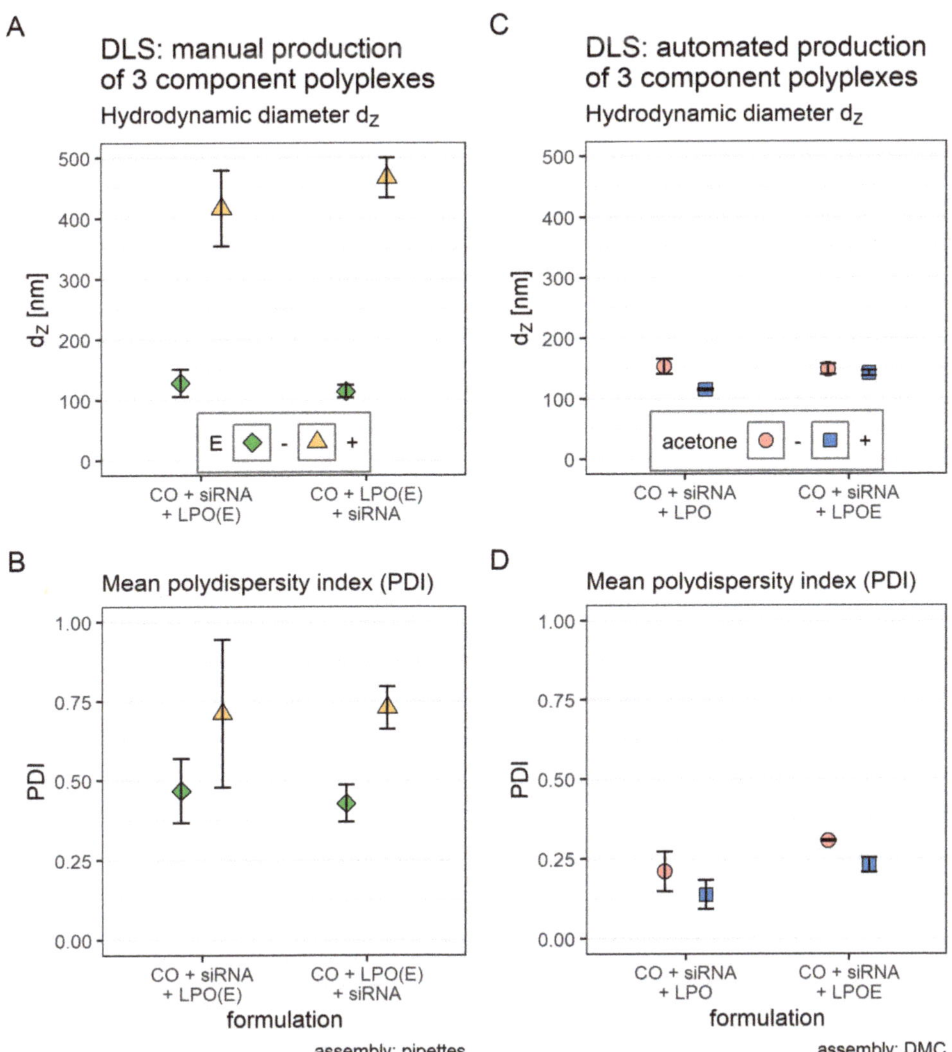

Figure 5. Manual or automated formulation of three component siRNA polyplexes. DLS data are represented as the mean of three measurements. Color and shape encode either the difference in the sequence of lipid anchored PEG12 oligomers (LPO, with or without E (glutamic acid), panel **A**,**B**) or the difference in formulation conditions (oligomer CO dissolved in HBG with or without 50% acetone, panel C,D). (**A**,**B**): polyplexes were formulated manually by mixing all educts with pipettes. The mixing order is denoted on the x-axis. Orange triangle: the sequence of the LPO contains two additional glutamic acids. Green diamond: no additional glutamic acids. (**C**,**D**): polyplexes were formulated automatically inside the double meander channel (DMC, Figure 1) by the nanoparticle production system. Flow rates: siRNA 900 μL/h (S4), CO 100 μL/h (S3). LPO(E) 50 μL/h (S2, two syringes). Total flow rate: 1100 μL/h. The educts are denoted on the x-axis. Red dots: CO was dissolved in HBG only. Blue squares: CO was dissolved in HBG with 50% acetone. (**A**,**C**): mean hydrodynamic diameter (z-average). (**B**,**D**): Mean polydispersity index (PDI). Error bars correspond to 95% confidence intervals. N = 3. Raw data were selected from our previous publication [30], here presented in a new format.

In Figure 5C,D, polyplexes from the three educts were produced automatically inside the double meander channel (DMC) by the nanoparticle production system. In the first mixing zone, CO and siRNA were mixed. The color and the shape of the data points indicate if CO was dissolved in HBG with or without 50% acetone. In the second mixing zone, LPO or LPOE was added to the mixture. Polyplexes prepared from CO dissolved in HBG only showed slightly higher hydrodynamic diameters d_Z and PDI (CO + siRNA + LPO: d_Z = 153.0 ± 12.7 nm, PDI = 0.210 ± 0.062; CO + siRNA + LPOE: d_Z = 148.2 ± 8.7 nm, PDI = 0.306 ± 0.003) in comparison to polyplexes prepared from CO dissolved in HBG with 50% acetone (CO + siRNA + LPO: d_Z = 114.7 ± 1.5 nm, PDI = 0.137 ± 0.045; CO + siRNA + LPOE: d_Z = 141.9 ± 4.7 nm, PDI = 0.230 ± 0.022).

Automated production of three component polyplexes (CO + siRNA + LPO) generated nanoparticles with smaller hydrodynamic diameters and PDIs compared to manually prepared polyplexes. Incorporating glutamic acid into the structure of the lipid anchor PEG12 oligomer facilitated the production of polyplexes with comparable mean hydrodynamic diameters regardless of production method. Nevertheless, the PDI of manually prepared polyplexes was still larger than the PDI of polyplexes prepared with the nanoparticle production system.

3. Discussion

A detailed description of the automated nanoparticle production system, its hardware, and its software is provided. The control module is inexpensive, and its parts are readily available. The feeding module integrates syringe pumps as commonly applied in microfluidic systems such as in Liu et al. (PHD 2000, Harvard Apparatus) [36], Debus et al. (Aladdin, World precision Instruments) [37], Lim et al. (model unspecified, Harvard Apparatus) [38], Karnik et al. (SP220I, World Precision Instruments and PHD 22/2000, Harvard Apparatus) [39], and Belliveau et al. (KD200, KD Scientific) [27]. The schematics of the collection module are published together with this paper on GitHub [32], enabling the replication of this module in any workshop. Additionally, building the collection module with additive fabrication methods, e.g., 3-D printing, might also be feasible.

The software that controls the feeding and the collection module enhances the functionality of any formulation module. With a specific, customized main module for each individual experiment, reproducibility is increased since every production cycle follows the same commands. Additionally, logs of each experiment are available to document the intended execution of the program. With each main module tailored to the specific needs of any experiment, repeating an experiment is done by simply executing the program again. Additional benefits of employing the software to control the feeding module are the ramping and purging functionalities that reduce the waste of educts to a minimum. These functionalities are cumbersome at best to program into each pump manually, but readily available in our software. The ramping functions ensure that all educts reach the mixing zone at the same time and the purging functions enable the user to choose the least expensive reagent to purge the product from the channel after the experiment.

Although volumes and flow rates of each experiment can be taken from its log, the components and concentrations must still be recorded manually. We have demonstrated the importance of detailed experiment descriptions with polyplexes prepared manually from two components. Changing the volume ratios of the educts and the mixing order varied the hydrodynamic diameter and PDI of the resulting polyplexes from 52.5 ± 9.7 nm to 141.0 ± 3.2 nm and 0.249 ± 0.144 to 0.109 ± 0.030, respectively. The effect of volume ratios on particle sizes is probably due to turbulences of varying intensity, which usually promote faster mixing of the educts. This effect is especially important during the polyplex complexation process since charge neutralization occurs in around 50 ms [21]. Additionally, we demonstrated that some formulations might be impossible to be produced with pipettes and require a formulation module—especially formulations from three or more components seem to benefit from the increased control of a microfluidic setup. Krzysztoń et al., for example, improved the efficiency of their mNALP (monomolecular nucleic acid/lipid particles) formulation by microfluidic mixing on the same hydrodynamic flow-focusing chip without our device [29]. We have prepared three component polyplexes manually and automatically. Polyplexes prepared by rapid pipetting showed hydrodynamic diameters and PDIs in suboptimal ranges. With the automated nanoparticle production system, polyplexes with d_Z = 114.7 ± 1.5 nm and PDI = 0.137 ± 0.045 could be produced. This finding and the application of the automated nanoparticle production system enabled the establishment of structure-function relationships from three component polyplexes in our previous publication [30].

Sizes and PDIs in a desired range, however, do not automatically guarantee superior biological activity of target nanoparticles in vitro or in vivo. On the one hand, nanoparticles produced with controlled methods might show improved formulation characteristics and equal (but not better) biological activity. Members from our lab, for example, demonstrated the reproducible production of polyplexes from pDNA and LPEI (linear polyethylene imine) with an up-scaled micro-mixer. Compared with manually formulated polyplexes, both formulations showed comparable biological activity in vitro [23]. On the other hand, formulations with larger PDIs and sizes might show apparently better transfection efficiencies in vitro. This is usually due to large particles literally "dropping" on the cells fixed to the bottom of the cell culture flask. A formulation with these properties, however, might fail in vivo.

All in all, the application of this versatile software enables the creation and automated execution of a sophisticated program consisting of many individual steps in order to increase control over the formulation process of nanoparticles and foster reproducibility, which will be most relevant for pharmaceutical production.

The next step on the course to automation is the integration of a fraction collector. The device developed here was designed to work with any standard well plate to realize product collection and separation. It is independent of the previously mentioned setup, which makes it suitable for a wide range of applications. It can be integrated into the target automated process, but it can also be used to gather products produced manually. Overall, it is a versatile addition to any product formulation setup relieving the user of additional manual labor.

A well-known disadvantage of microfluidic systems is the scalability problem [24]. Due to the utilization of fluid phenomena—for example, laminar flow—which are only present under certain conditions, the throughput of one microfluidic chip cannot be escalated indefinitely [25]. The obvious solution to employ parallelization is a valid suggestion, but product output does only scale linearly in relation to dedicated resources at the current development stage of the system due to the many individual steps involved in setting up the device. However, a possible solution is already designed in the system. Since the setup is modular, any part of it can easily be replaced by a more efficient one [40].

Placing the complete system into a laminar flow cabinet is a next obvious step for pharmaceutical applications. In the described work, nanoparticles were formulated outside of the cabinet and subsequently transferred inside for in vitro transfections of cells. With the complete system inside the cabinet, direct application of the product to the target cells could be achieved, which would decrease the influences of external factors and human interactions even further.

4. Materials and Methods

4.1. Materials

The materials and the software used for the control module are listed in the Supplemental Information (2.1. Materials, Table S2; 2.2. Software, Table S3).

The materials used for building the feeding module are listed in the Supplemental Information (3.1. Materials, Table S4).

Materials used for building the formulation modules used in this publication are listed in the Supplemental Information (1.1. Materials, Table S1).

The design and the manufacturing procedure of the microfluidic chips is described in Krzysztoń et al., 2017 [29].

In brief, the design of the microfluidic channels was realized on a silica wafer with soft lithographic methods. The finished wafer was covered with polydimethylsiloxane

(PDMS) mixed with 10% (*w*/*w*) crosslinker, degassed, and cured (75 °C, 4 h). Subsequently, the solid PDMS was cut and removed from the wafer. Inlets and outlets were pierced with a biopsy puncher and the channel was bonded to a glass slide with an oxygen plasma cleaner (Diener Electronic; 10 W high frequency generator power, 12 s, Pico Model E). Polyethylene tubes (length: 110 mm, inner diameter 0.38 mm) were fitted to the holes and the complete chip was covered in another layer of PDMS to increase resistance to pressure. Wang et al., 2014 investigated the tubing of PDMS channels and offer an improved protocol to prevent channel leakage [33]. Each new channel was tested before application with a standard formulation. The size and PDI measured by DLS were compared to the results from the same formulation produced with the previous channel. A to-scale model of the channels can be found in the Supplemental Information (1.2. Schematics, Figures S1 and S2). We calculated the Reynold's number (Re), Dean's number (De), and backpressure (ΔP) for the SMC and DMC at a total flow rate of 1500 µL/h: SMC: Re $\approx$ 3, De $\approx$ 1.23, ΔP = 1249.4 mbar; DMC: Re $\approx$ 2.5, De $\approx$ 1.25, ΔP = 2498.8 mbar. Solvents used in this paper are classified as low-solubility solvents which are compatible with microfluidic systems fabricated in PDMS by Lee et al. Therefore, they are unlikely to cause considerable changes to the channel geometry due to swelling [41].

Materials used for building the collection module are listed in the Supplemental Information (4.1. Materials, Table S5).

A prototype of the collection module (fraction collector) was built according to the design published at GitHub (https://github.com/Dominikmloy/fraction-collector.git). All parts were cut from aluminum, except the parts noted below. The prototype was built by the workshop of the LMU Munich.

Materials for the formulation of polyplexes and DLS measurements: The synthesis of the oligomers (CO: id = 991, LPO: id = 1203, LPOE: id = 1223) and the required materials are described in detail in [30,31]. The sequences of all oligomers are depicted in the Supplemental Information (5. Oligomers, Figure S7). Solvents: Purified water (produced with Ultra Clear GP UV UF, Evoqua Water Technologies GmbH, Günzburg, Germany), acetone HPLC grade (VWR international GmbH, Darmstadt, Germany). Chemicals: 4-(2-hydroxyethyl)-1-piperazineethanesulfonic acid ultra-pure (HEPES, Biomol GmbH, Hamburg, Germany), D(+)glucose monohydrate DAB (Loewe Biochemica GmbH, Sauerlach, Germany), NaOH pellets puriss. (VWR international GmbH, Darmstadt, Germany), NaOH 1M standard solution (Thermo Fisher Scientific GmbH, Schwerte, Germany), HCl 1M standard solution (VWR international GmbH, Darmstadt, Germany). Nucleic acids: siGFP sense: 5′-AuAucAuGGccGAcAAGcAdTsdT-3′, antisense: 5′-UGCUUGUCGGCcAUGAuAUdTsdT-3′ (Axolabs, Kulmbach, Germany). Small letters: 2′ methoxy; s: phosphorothioate.

4.2. Methods

4.2.1. Polyplex Preparation

Polyplexes were prepared with a final siRNA concentration of 0.025 mg/mL. A nitrogen to phosphate (N/P) ratio of 12 was used to determine the amount of core oligomer CO relative to the amount of siRNA. The N/P ratio relates the number of positive charges from the primary and secondary amines in the oligomer's backbone to the number of negative charges from the phosphates in the siRNA's backbone. The manual method of polyplex preparation was done with pipettes and rapid mixing in a batch wise process. The solvent—if not noted differently—was HEPES buffer pH 7.4 with 5% glucose (HBG). This buffer was used because it does not rely on salts to be isotonic, since polyplex formation relies on charge interactions that could be hampered by ions.

Manual polyplex preparation: CO solution (0.504 mg/mL) was added quickly to a siRNA solution (0.05 mg/mL) of equal volume and mixed by rapid pipetting, achieving a final siRNA concentration of 0.025 mg/mL. Subsequently, the formulation was incubated for 45 min. Concentrations and volumes for mixing polyplexes from unequal volumes were adjusted accordingly: 5.8 µL of CO at 3.023 mg/mL, or 64.2 µL of CO at 0.275 mg/mL. 64.2 µL of siRNA at 0.027 mg/mL, or 5.8 µL of siRNA at 0.300 mg/mL.

For the manual formulation of three component polyplexes, equal volumes (27.7 µL) of CO solution (0.637 mg/mL) and siRNA solution (0.063 mg/mL) were used. The amount of LPO and LPOE was set to 20 mol% relative to CO. Concentrations were set to 0.207 mg/mL LPO, 0.224 mg/mL LPOE, volumes were 14.6 µL. Solutions were mixed sequentially by rapid pipetting. When siRNA was used in the first step, a ten-minute break was taken after the two components were mixed to allow the polyplex to stabilize. After the addition of the third component, the formulation was incubated for 45 min before DLS was measured.

Automated polyplex preparation: The formulation module with the double meander channel (DMC) was used without any additional surface treatment (Figure 1). Before each usage, the channel was washed and primed with the same solvents that were used to produce the polyplexes. Details about the washing/priming process can be found in the Supplemental Information (3.2.4 Module: setup.py). siRNA in HBG (0.033 mg/mL) was loaded into S4 (FR = 900 µL/h) and CO (3.025 mg/mL) in HBG or HBG with 50% acetone to retard siRNA compaction was loaded into S3 (FR = 100 µL/h). LPO or LPOE in HBG with 50% acetone to facilitate solvent exchange were loaded into S2. The flow rate of each syringe S2 was 50 µL/h at a total flow rate of 1100 µL/h, resulting in a flow rate ratio of lipid anchor oligomer to core polyplex of 1:11. The final product was diluted with HBG to 0.025 mg siRNA/mL.

Stability of the formulation presented in this paper has been investigated previously. Troiber et al. have found particles assembled from the same class of oligomers by rapid pipetting to be stable over three weeks [42]. In our previous paper, we have investigated the changes in size, PDI, and zeta potential of our core formulation (siRNA and CO) over 90 min. The core formulation was assembled in the single meander channel (SMC) [30]. We saw no changes in size and PDI. However, changes in the zeta potential of the particles up to the 40 min mark were the reason why formulations were always used after 45 min incubation time.

4.2.2. DLS Measurement

Samples used for dynamic light scattering (DLS) measurements were prepared to contain 1.5 µg siRNA in 60 µL HEPES buffered glucose pH 7.4 (HBG) at 25 °C and the corresponding amount of oligomer. Refractive index and viscosity of the solution were calculated using the solvent builder integrated into the software (Zetasizer family software update v7.12). Viscosities and refractive indices (RI) are reported in Figure S6 in the Supplemental Information. RI of all particles was estimated to be 1.45. Scattered light used to determine the hydrodynamic diameter of polyplexes was measured at a 173° angle (backscatter) with a flexible attenuator with a Zetasizer Nano ZS ZEN 3600 (Malvern Panalytical Ltd., Malvern, UK) in DTS1070 micro cuvettes (Malvern Panalytical Ltd., Malvern, UK). Samples were measured three times with 12–15 sub runs each. The mean z-average in nm of those three runs is reported with error bars corresponding to the 95% confidence interval of the three runs.

4.2.3. Standardization of the System

The following steps were taken to ensure standardization of the system.

1. Microfluidic channels were always prepared from the same silica wafer template.
2. Once the optimal formulation conditions for a target formulation were established, the respective mixing program was stored on the raspberry pi.
3. Each formulation produced with this system is measured by DLS.
4. Changes made to the system are validated with a standard formulation.

4.2.4. Data Analysis

Data was analyzed with R [43] and RStudio [44]. We always report means with 95% confidence intervals. R code and raw data are made available here: DOI: 10.6084/m9.figshare.13285577.

Supplementary Materials: The following are available online at https://www.mdpi.com/2227-9717/9/1/129/s1: Figure S1: Channel design of the single meander channel, Figure S2: Channel design of the double meander channel, Figure S3: UML class diagram of the control software of the syringe pumps, Figure S4: Flowchart describing the automation process of polyplex formulations, Figure S5: UML class diagram of the fraction collector's control software, Figure S6: Flowchart describing the workflow of the fraction collector, Figure S7: Chemical structures of CO, LPO, LPOE. Table S1: Materials formulation module, Table S2: Materials control module, Table S3: Software control module, Table S4: Materials feeding module, Table S5. Materials collection module, Table S6. Solvents used for DLS measurements. Code 1: Excerpt from the '_set_from_spec_file()' function, Code 2: Excerpt from the initialization of the logger function, Code 4: 'GlobalPhaseNumber()' class, Code 5: Excerpt from the 'Setup()' class, Code 6: Excerpt from the 'check_connections()' function, Code 7: Excerpt from the 'rate()' function, Code 8: Excerpt from the 'ramping_calc()' function, Code 9: Excerpt from the 'overlap_calc()' function, Code 10: 'main.py' module, Code 11: Excerpt from the 'Initialize()' class, Code 12: Excerpt from the 'Move()' class, Code 13: The 'main.py' module of the fraction collector.

Author Contributions: Conceptualization, D.M.L., R.K., U.L., J.O.R., and E.W.; data curation, D.M.L.; formal analysis, D.M.L.; funding acquisition, J.O.R. and E.W.; investigation, D.M.L.; methodology, D.M.L., R.K., U.L., J.O.R., and E.W.; project administration, D.M.L.; resources, D.M.L., R.K., U.L., J.O.R., and E.W.; software, D.M.L.; supervision, E.W.; validation, D.M.L., U.L. and E.W.; visualization, D.M.L.; writing—original draft, D.M.L.; writing—review and editing, D.M.L., R.K., and E.W. All authors have read and agreed to the published version of the manuscript.

Funding: This work was supported by the Deutsche Forschungsgemeinschaft (DFG), 201269156-SFB 1032 (projects B1 Rädler and B4 Wagner), the Munich Center for NanoScience (CeNS), and the Cluster of Excellence Nanosystems Initiative Munich (NIM). Rafał Krzysztoń was supported by German Research Foundation (DFG) through the Graduate School of Quantitative Biosciences Munich (QBM) (GSC 1006). The funders had no role in study design, data collection and analysis, decision to publish, or preparation of the manuscript.

Institutional Review Board Statement: Not applicable.

Informed Consent Statement: Not applicable.

Data Availability Statement: The data presented in this study are openly available in FigShare at DOI: 10.6084/m9.figshare.13285577 [45].

Acknowledgments: We thank Wolfgang Rödl for technical support and we thank the workshop of the LMU for their enormous help in building the collection module. We are grateful to Philipp Klein for synthesis of the CO oligomer. We thank Thomas Unterlinner for his patience and proficiency in answering all our questions about stepper motors.

Conflicts of Interest: The authors declare no conflict of interest.

References

1. Li, S.-D.; Huang, L. Pharmacokinetics and Biodistribution of Nanoparticles. *Mol. Pharm.* **2008**, *5*, 496–504. [CrossRef] [PubMed]
2. Li, M.; Al-Jamal, K.T.; Kostarelos, K.; Reineke, J. Physiologically Based Pharmacokinetic Modeling of Nanoparticles. *ACS Nano* **2010**, *4*, 6303–6317. [CrossRef] [PubMed]
3. Moghimi, S.M.; Hunter, A.C.; Andresen, T.L. Factors Controlling Nanoparticle Pharmacokinetics: An Integrated Analysis and Perspective. *Annu. Rev. Pharmacol. Toxicol.* **2012**, *52*, 481–503. [CrossRef] [PubMed]
4. Gradishar, W.J.; Tjulandin, S.; Davidson, N.; Shaw, H.; Desai, N.; Bhar, P.; Hawkins, M.; O'Shaughnessy, J. Phase III Trial of Nanoparticle Albumin-Bound Paclitaxel Compared with Polyethylated Castor Oil—Based Paclitaxel in Women with Breast Cancer. *J. Clin. Oncol.* **2005**, *23*, 7794–7803. [CrossRef] [PubMed]
5. Dewolf, H.; Snel, C.; Verbaan, F.; Schiffelers, R.; Hennink, W.; Storm, G. Effect of cationic carriers on the pharmacokinetics and tumor localization of nucleic acids after intravenous administration. *Int. J. Pharm.* **2007**, *331*, 167–175. [CrossRef] [PubMed]
6. Zintchenko, A.; Susha, A.S.; Concia, M.; Feldmann, J.; Wagner, E.; Rogach, A.L.; Ogris, M. Drug Nanocarriers Labeled with Near-infrared-emitting Quantum Dots (Quantoplexes): Imaging Fast Dynamics of Distribution in Living Animals. *Mol. Ther.* **2009**, *17*, 1849–1856. [CrossRef] [PubMed]
7. Jang, J.; Oh, J.H. A top-down approach to fullerene fabrication using a polymer nanoparticle precursor. *Adv. Mater.* **2004**, *16*, 1650–1653. [CrossRef]

8. Rolland, J.P.; Maynor, B.W.; Euliss, L.E.; Exner, A.E.; Denison, G.M.; DeSimone, J.M. Direct fabrication and harvesting of monodisperse, shape-specific nanobiomaterials. *J. Am. Chem. Soc.* **2005**, *127*, 10096–10100. [CrossRef]
9. Sun, W.; Mao, S.; Shi, Y.; Li, L.C.; Fang, L. Nanonization of itraconazole by high pressure homogenization: Stabilizer optimization and effect of particle size on oral absorption. *J. Pharm. Sci.* **2011**, *100*, 3365–3373. [CrossRef]
10. Hu, C.M.J.; Zhang, L.; Aryal, S.; Cheung, C.; Fang, R.H.; Zhang, L. Erythrocyte membrane-camouflaged polymeric nanoparticles as a biomimetic delivery platform. *Proc. Natl. Acad. Sci. USA* **2011**, *108*, 10980–10985. [CrossRef]
11. Chen, Y.; Zhou, H.; Wang, Y.; Li, W.; Chen, J.; Lin, Q.; Yu, C. Substrate hydrolysis triggered formation of fluorescent gold nanoclusters-a new platform for the sensing of enzyme activity. *Chem. Commun.* **2013**, *49*, 9821–9823. [CrossRef] [PubMed]
12. Chan, H.-K.; Kwok, P.C.L. Production methods for nanodrug particles using the bottom-up approach. *Adv. Drug Deliv. Rev.* **2011**, *63*, 406–416. [CrossRef] [PubMed]
13. Lächelt, U.; Wagner, E. Nucleic Acid Therapeutics Using Polyplexes: A Journey of 50 Years (and Beyond). *Chem. Rev.* **2015**, *115*, 11043–11078. [CrossRef] [PubMed]
14. Felgner, P.L.; Barenholz, Y.; Behr, J.P.; Cheng, S.H.; Cullis, P.; Huang, L.; Jessee, J.A.; Seymour, L.; Szoka, F.; Thierry, A.R.; et al. Nomenclature for Synthetic Gene Delivery Systems. *Hum. Gene Ther.* **1997**, *8*, 511–512. [CrossRef] [PubMed]
15. Kabanov, A.V.; Kabanov, V.A. DNA Complexes with Polycations for the Delivery of Genetic Material into Cells. *Bioconjug. Chem.* **1995**, *6*, 7–20. [CrossRef]
16. Merrifield, R.B. Solid Phase Peptide Synthesis. I. The Synthesis of a Tetrapeptide. *J. Am. Chem. Soc.* **1963**, *85*, 2149–2154. [CrossRef]
17. Schaffert, D.; Badgujar, N.; Wagner, E. Novel Fmoc-polyamino acids for solid-phase synthesis of defined polyamidoamines. *Org. Lett.* **2011**, *13*, 1586–1589. [CrossRef]
18. Schaffert, D.; Troiber, C.; Salcher, E.E.; Fröhlich, T.; Martin, I.; Badgujar, N.; Dohmen, C.; Edinger, D.; Kläger, R.; Maiwald, G.; et al. Solid-phase synthesis of sequence-defined T-, i-, and U-shape polymers for pDNA and siRNA delivery. *Angew. Chem. Int. Ed. Engl.* **2011**, *50*, 8986–8989. [CrossRef]
19. Scholz, C.; Kos, P.; Wagner, E. Comb-like oligoaminoethane carriers: Change in topology improves pDNA delivery. *Bioconjug. Chem.* **2014**, *25*, 251–261. [CrossRef]
20. Ou, Z.; Muthukumar, M. Entropy and enthalpy of polyelectrolyte complexation: Langevin dynamics simulations. *J. Chem. Phys.* **2006**, *124*, 154902. [CrossRef]
21. Braun, C.S.; Fisher, M.T.; Tomalia, D.A.; Koe, G.S.; Koe, J.G.; Middaugh, C.R. A stopped-flow kinetic study of the assembly of nonviral gene delivery complexes. *Biophys. J.* **2005**, *88*, 4146–4158. [CrossRef] [PubMed]
22. Zelphati, O.; Nguyen, C.; Ferrari, M.; Felgner, J.; Tsai, Y.; Felgner, P.L. Stable and monodisperse lipoplex formulations for gene delivery. *Gene Ther.* **1998**, *5*, 1272–1282. [CrossRef] [PubMed]
23. Kasper, J.C.; Schaffert, D.; Ogris, M.; Wagner, E.; Friess, W. The establishment of an up-scaled micro-mixer method allows the standardized and reproducible preparation of well-defined plasmid/LPEI polyplexes. *Eur. J. Pharm. Biopharm.* **2011**, *77*, 182–185. [CrossRef] [PubMed]
24. Whitesides, G.M. The origins and the future of microfluidics. *Nature* **2006**, *442*, 368–373. [CrossRef] [PubMed]
25. Squires, T.M.; Quake, S.R. Microfluidics: Fluid physics at the nanoliter scale. *Rev. Mod. Phys.* **2005**, *77*, 977–1026. [CrossRef]
26. Koh, C.G.; Kang, X.; Xie, Y.; Fei, Z.; Guan, J.; Yu, B.; Zhang, X.; Lee, L.J. Delivery of Polyethylenimine/DNA Complexes Assembled in a Microfluidics Device. *Mol. Pharm.* **2009**, *6*, 1333–1342. [CrossRef]
27. Belliveau, N.M.; Huft, J.; Lin, P.J.; Chen, S.; Leung, A.K.; Leaver, T.J.; Wild, A.W.; Lee, J.B.; Taylor, R.J.; Tam, Y.K.; et al. Microfluidic Synthesis of Highly Potent Limit-size Lipid Nanoparticles for In Vivo Delivery of siRNA. *Mol. Ther. Nucleic Acids* **2012**, *1*, e37. [CrossRef]
28. Grigsby, C.L.; Ho, Y.-P.; Lin, C.; Engbersen, J.F.J.; Leong, K.W. Microfluidic Preparation of Polymer-Nucleic Acid Nanocomplexes Improves Nonviral Gene Transfer. *Sci. Rep.* **2013**, *3*, 3155. [CrossRef]
29. Krzysztoń, R.; Salem, B.; Lee, D.J.; Schwake, G.; Wagner, E.; Rädler, J.O. Microfluidic self-assembly of folate-targeted monomolecular siRNA-lipid nanoparticles. *Nanoscale* **2017**, *9*, 7442–7453. [CrossRef]
30. Loy, D.M.; Klein, P.M.; Krzysztoń, R.; Lächelt, U.; Rädler, J.O.; Wagner, E. A microfluidic approach for sequential assembly of siRNA polyplexes with a defined structure-activity relationship. *PeerJ Mater. Sci.* **2019**, *1*, e1. [CrossRef]
31. Klein, P.M.; Kern, S.; Lee, D.-J.; Schmaus, J.; Höhn, M.; Gorges, J.; Kazmaier, U.; Wagner, E. Folate receptor-directed orthogonal click-functionalization of siRNA lipopolyplexes for tumor cell killing in vivo. *Biomaterials* **2018**, *178*, 630–642. [CrossRef] [PubMed]
32. Loy, D.M. Dominik Loy on GitHub. 2020. Available online: https://github.com/Dominikmloy (accessed on 7 November 2020).
33. Wang, J.; Chen, W.; Sun, J.; Liu, C.; Yin, Q.; Zhang, L.; Xianyu, Y.; Shi, X.; Hu, G.; Jiang, X. A microfluidic tubing method and its application for controlled synthesis of polymeric nanoparticles. *Lab Chip* **2014**, *14*, 1673–1677. [CrossRef] [PubMed]
34. Andersen, G. XY Table. 2016. Available online: https://www.thingiverse.com/thing:18678 (accessed on 18 January 2020).
35. Sletten, E.M.; Bertozzi, C.R. From mechanism to mouse: A tale of two bioorthogonal reactions. *Acc. Chem. Res.* **2011**, *44*, 666–676. [CrossRef] [PubMed]
36. Liu, D.; Cito, S.; Zhang, Y.; Wang, C.-F.; Sikanen, T.M.; Santos, H.A. A Versatile and Robust Microfluidic Platform Toward High Throughput Synthesis of Homogeneous Nanoparticles with Tunable Properties. *Adv. Mater.* **2015**, *27*, 2298–2304. [CrossRef]
37. Debus, H.; Beck-Broichsitter, M.; Kissel, T. Optimized preparation of pDNA/poly(ethylene imine) polyplexes using a microfluidic system. *Lab Chip* **2012**, *12*, 2498. [CrossRef]

38. Lim, J.-M.; Swami, A.; Gilson, L.M.; Chopra, S.; Choi, S.; Wu, J.; Langer, R.; Karnik, R.; Farokhzad, O.C. Ultra-High Throughput Synthesis of Nanoparticles with Homogeneous Size Distribution Using a Coaxial Turbulent Jet Mixer. *ACS Nano* **2014**, *8*, 6056–6065. [CrossRef]
39. Karnik, R.; Gu, F.; Basto, P.; Cannizzaro, C.; Dean, L.; Kyei-Manu, W.; Langer, R.; Farokhzad, O.C. Microfluidic Platform for Controlled Synthesis of Polymeric Nanoparticles. *Nano Lett.* **2008**, *8*, 2906–2912. [CrossRef]
40. Chiu, D.T.; DeMello, A.J.; Di Carlo, D.; Doyle, P.S.; Hansen, C.; Maceiczyk, R.M.; Wootton, R.C.R. Small but Perfectly Formed? Successes, Challenges, and Opportunities for Microfluidics in the Chemical and Biological Sciences. *Chemistry* **2017**, *2*, 201–223. [CrossRef]
41. Lee, J.N.; Park, C.; Whitesides, G.M. Whitesides Solvent compatibility of poly (dimethylsiloxane)-based microfluidic devices. *Anal. Chem.* **2003**, *75*, 6544–6554. [CrossRef]
42. Troiber, C.; Kasper, J.C.; Milani, S.; Scheible, M.; Martin, I.; Schaubhut, F.; Küchler, S.; Rädler, J.; Simmel, F.C.; Friess, W.; et al. Comparison of four different particle sizing methods for siRNA polyplex characterization. *Eur. J. Pharm. Biopharm.* **2013**, *84*, 255–264. [CrossRef]
43. R Core Team. *R: A Language and Environment for Statistical Computing*; R Foundation for Statistical Computing: Vienna, Austria, 2018.
44. RStudio Team. *RStudio: Integrated Development for R*; RStudio, Inc.: Boston, MA, USA, 2018.
45. Loy, D.M. Controlling Nanoparticle Formulation: A Low-Budget Prototype for the Automation of a Microfluidic Platform. 2021. Available online: https://figshare.com/articles/dataset/Controlling_nanoparticle_formulation_a_low-budget_prototype_for_the_automation_of_a_microfluidic_platform/13285577 (accessed on 7 January 2021). [CrossRef]

 MDPI

Article

Centrifugal Microfluidic Integration of 4-Plex ddPCR Demonstrated by the Quantification of Cancer-Associated Point Mutations

Franziska Schlenker [1], Elena Kipf [1], Nadine Borst [1,2], Nils Paust [1,2], Roland Zengerle [1,2], Felix von Stetten [1,2], Peter Juelg [1,†] and Tobias Hutzenlaub [1,2,*,†]

[1] Hahn-Schickard, Georges-Koehler-Allee 103, 79110 Freiburg, Germany; Franziska.schlenker@hahn-schickard.de (F.S.); elena.kipf@hahn-schickard.de (E.K.); Nadine.Borst@Hahn-Schickard.de (N.B.); Nils.Paust@hahn-schickard.de (N.P.); Roland.Zengerle@Hahn-Schickard.de (R.Z.); Felix.von.Stetten@Hahn-Schickard.de (F.v.S.); Peter.Juelg@Hahn-Schickard.de (P.J.)

[2] Laboratory for MEMS Applications, IMTEK—Department of Microsystems Engineering, University of Freiburg, Georges-Koehler-Allee 103, 79110 Freiburg, Germany

* Correspondence: Tobias.Hutzenlaub@Hahn-Schickard.de; Tel.: +49-761-203-73269

† Contributed equally.

Abstract: We present the centrifugal microfluidic implementation of a four-plex digital droplet polymerase chain reaction (ddPCR). The platform features 12 identical ddPCR units on a LabDisk cartridge, each capable of generating droplets with a diameter of 82.7 ± 9 µm. By investigating different oil–surfactant concentrations, we identified a robust process for droplet generation and stabilization. We observed high droplet stability during thermocycling and endpoint fluorescence imaging, as is required for ddPCRs. Furthermore, we introduce an automated process for four-color fluorescence imaging using a commercial cell analysis microscope, including a customized software pipeline for ddPCR image evaluation. The applicability of ddPCRs is demonstrated by the quantification of three cancer-associated *KRAS* point mutations (G12D, G12V and G12A) in a diagnostically relevant wild type DNA background. The four-plex assay showed high sensitivity (3.5–35 mutant DNA copies in 15,000 wild type DNA copies) and linear performance (R^2 = 0.99) across all targets in the LabDisk.

Keywords: microfluidics; digital droplet polymerase chain reaction (ddPCR); multiplexing; centrifugal step emulsification; droplet stability; droplet fluorescence evaluation

Citation: Schlenker, F.; Kipf, E.; Borst, N.; Paust, N.; Zengerle, R.; von Stetten, F.; Juelg, P.; Hutzenlaub, T. Centrifugal Microfluidic Integration of 4-Plex ddPCR Demonstrated by the Quantification of Cancer-Associated Point Mutations. *Processes* **2021**, *9*, 97. https://doi.org/10.3390/pr9010097

Received: 30 November 2020
Accepted: 23 December 2020
Published: 5 January 2021

1. Introduction

Digital polymerase chain reaction (dPCR) allows the precise and absolute quantification of nucleic acids without the need for standard curves [1,2]. Due to its sensitivity, dPCR is particularly useful for detecting low target concentrations in complex backgrounds, such as rare point mutations in a background of non-mutated wild type DNA [3]. In contrast to conventional quantitative PCR, the reaction mix is partitioned into thousands of compartments using either pre-defined stationary cavities or aqueous droplets in oil [4–7]. In recent years, the latter approach, digital droplet PCR (ddPCR), has become the most common, with companies such as Bio-Rad and Stilla Technologies releasing commercial systems [8]. One of the key factors for a successful ddPCR is the ability to generate emulsions that are stable throughout the complete analytic workflow. This is especially challenging during DNA amplification, when temperature cycling induces thermal stress [9]. The most basic approach is to stabilize these emulsions with surfactants in the oil phase. However, this is typically not sufficient, and further measures of droplet stabilization have been introduced. Bio-Rad additionally uses reagents which encapsulate the droplets with a protein skin by applying heat. However, this has the downside that the customer is limited to a specific

PCR mix, obtained from Bio-Rad [10,11]. Stilla Technologies generates and cycles the droplets in their device by applying up to 950 mbar of overpressure to stabilize them. This requires an external pressure source that is not available in all settings (e.g., at the point of care (PoC)) [12].

In this work, we investigate the influence of surfactant concentration and oil combinations for a ddPCR without any additional means of stabilization. Furthermore, we introduce a simple and robust laboratory workflow for image analysis of a four-color ddPCR performed inside a microfluidic cyclo-olefin copolymer (COC) cartridge. Our work is based on centrifugal microfluidic step emulsification, as introduced by Schuler et al. [9]. In the recent past, centrifugal microfluidics developed into a mature technology which enabled the efficient miniaturization, parallelization and integration of assays for both sample preparation and analysis [13–15]. In the past, several groups have presented centrifugal, microfluidic cartridges that were able to emulsify a reaction mix [6,9,16–20]. We present a centrifugal, microfluidic, single-use cartridge (LabDisk) with 12 identical ddPCR units. In contrast to the state of the art, where until now only one-plex ddPCR assays have been automated by centrifugal microfluidics [9,17–19], we present an increased multiplexing degree of four. This elevated degree of multiplexing is particularly relevant for cancer diagnostics, where a comprehensive overview of the mutation status of a patient must be generated from a limited sample amount [3]. We demonstrate this high degree of multiplexing for point mutations in colorectal carcinoma. In one reaction, we quantify three *KRAS* mutations (G12D, G12V and G12A) in a background of their corresponding wild type sequence. By simultaneously quantifying the wild type, one can determine diagnostically relevant mutant to wild type allele frequencies. In proof-of-concept experiments with serial dilutions of a DNA template, we investigate the assay linearity and sensitivity for all targets.

2. Materials and Methods

2.1. LabDisk Design and Manufacturing

The computer-aided design layout of the ddPCR unit (see Figure 1A) and the final LabDisk with 12 of these ddPCR units were designed using Solidworks (Dassault Systèmes). Each ddPCR unit included an inlet to transfer the reagents into the reaction chamber. A supply channel (width of 40 µm and depth of 40 µm) connected the inlet and the reaction chamber. The distance between the sealing foil and the bottom of the chamber was defined by a pyramidal structure [9] with a minimum depth of 80 µm and a maximum depth of 160 µm. The supporting structures (pillars) with a diameter of 500 µm ensured this distance. As indicated in Figure 1A, four identical nozzles, each with a width of 24 µm and a depth of 15 µm, were used for parallel droplet generation. The ddPCR unit was vented by a hole on the far side of the LabDisk.

The Hahn-Schickard foundry service manufactured the LabDisks. The fabrication of the master tool started with precision milling of polymethyl methacrylate (PMMA), followed by soft lithography to create the flexible polydimethylsiloxane (PDMS) master. The manufacturing process included the micro-thermoforming and thermal sealing of the cartridges [21]. The 300 µm COC foils (COC8007/COC6013) from Tekniplex were thermoformed using the PDMS master tool. These structured foils were thermally sealed by 200 µm foil (COC8007/COC6013, Tekniplex). Polytetrafluoroethylene (PTFE) membranes (PMV15N, POREX [22]) on the far side of the LabDisk closed the venting holes to prevent evaporation during thermocycling. After reagent transfer, the inlet holes were closed by a pressure-sensitive adhesive foil (9795R from 3M). To improve the distribution of hot air, a 1 mm thick cover (PMMA, Evonik) with cutouts between each ddPCR unit was put on top of the disk before thermocycling (Figure S5, see electronic supplementary information (ESI)).

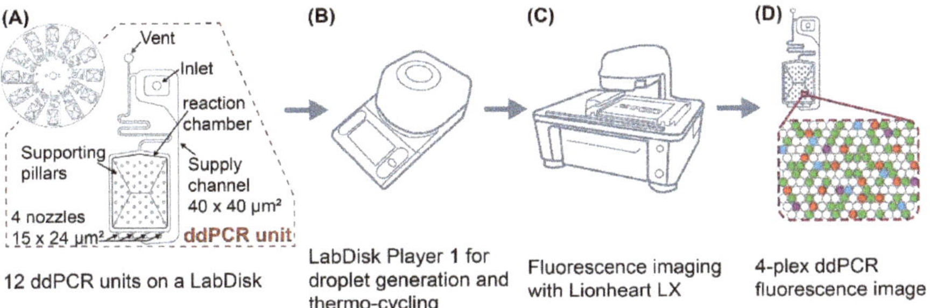

Figure 1. Workflow for an automated four-plex digital droplet polymerase chain reaction (ddPCR) in a LabDisk. (**A**) LabDisk with 12 identical ddPCR units. Each unit consisted of an inlet, a venting structure, a supply channel, a reaction chamber with supporting pillars and four nozzles for droplet generation by centrifugal step emulsification. (**B**) Droplet generation and thermocycling in LabDisk Player 1. (**C**) Fluorescence imaging with an automated Lionheart LX fluorescence microscope. (**D**) Four-plex ddPCR fluorescence image and data evaluation with Gen5 software to evaluate the assay performance.

2.2. PCR Reagents

For proof of concept, a Mediator Probe PCR was employed [23]. This assay principle is of particular interest for point mutations as well as for multiplexing, as the fluorescence signal generation at the universal reporter oligonucleotide is independent from the target sequence detected by label-free mediator probes [24]. The components of the four-plex mediator probe ddPCR mix were as follows: 1× concentrated PerfeCTa MultiPlex qPCR ToughMix (Quantabio), 1200 nM primers (biomers.net) and mediator probes (biomers.net) [24], 600 nM universal reporters (biomers.net) [24], HaeIII (NEB) digested human genomic DNA (Roche), and gBlock DNA fragments (131 bp, IDT). The preparation of the oligonucleotides and gBlocks and the DNA digestion were performed according to the manufacturers' manuals (see ESI). The 10 µL reaction mix for each dilution step contained 1 µL of wild type DNA and 1 µL of the mutant DNA target, both with a predefined concentration set by a 1× concentrated TE buffer (Tris-EDTA buffer solution T9285-100ML, Sigma-Aldrich) and nuclease-free water (AM9937, Ambion). In the proof of concept dilution series experiment, the four nominal concentrations of mutant DNA were 3540, 354, 35 and 3.5 copies per reaction for each mutant target, whereas the wild type DNA concentration was fixed at approximately 15,000 copies per reaction.

2.3. Droplet Generation and Thermocycling

To guarantee stable droplets during droplet generation and PCR thermocycling, the surfactant concentration and the oil combination must be chosen carefully. Earlier work [9] has already revealed the need for two different types of oil: (1) the emulsification-oil for homogeneous droplet generation and (2) the PCR-oil for droplet stabilization during PCR. In this work, the surfactant (DR-RE-SU-4G, Fluigent [25]) concentrations in the emulsification-oil (Novec7500, 3M [26]) and in the PCR-oil (FC-40, 3M [27]) were optimized.

Droplet generation and thermo cycling was performed in LabDisk Player 1 (DIALUNOX, formerly Qiagen Lake Constance). This PoC player (as illustrated in Figure 1B) is a small benchtop device for centrifugal actuation and temperature control. The complete frequency and temperature protocol for droplet cycling and the ddPCR is depicted in Figure 2. The transfer of the emulsification-oil to the reaction chamber was done at 40 Hz (120 s, 5 Hz/s). After pipetting 10 µL of the reaction mix into the inlet, the droplets (approximately 35,000) were generated by centrifugal step emulsification at 60 Hz for 3 min. It is important to accelerate to 60 Hz rapidly (15 Hz/s) in order to generate the droplets at a constant frequency. After droplet generation, the player was stopped (deceleration of 1 Hz/s), and the PCR-oil was pipetted into the inlet. The PCR-oil was transferred at 30 Hz over 10 min, with a very slow acceleration and deceleration rate of 1 Hz/s to avoid droplet merging. During droplet generation, the temperature remained constant at 25 °C. The ddPCR began with a hot start at 94 °C for 5 min, followed by 45 cycles between 94 °C (15 s) and 56 °C (60 s) and a 10 min cooling step at 25 °C at the end of the cycling protocol. During ddPCR thermocycling, the rotation frequency remained constant at 1 Hz to keep the droplets in the reaction chamber. After PCR thermocycling, the temperature was kept constant at 25 °C to allow the temperature in the ddPCR unit to reach room temperature. As we observed the post-PCR merging of droplets, most likely caused by electrostatic discharge effects, a standard handling procedure was developed: we recommend electrostatic discharge of the user prior to the transfer of the disk from LabDisk Player 1 to a lint-free tissue sprinkled with an antistatic spray (Screen Clene, Advanced Technology Cleaning). Using this procedure, we did not observe any further post-PCR merging effects.

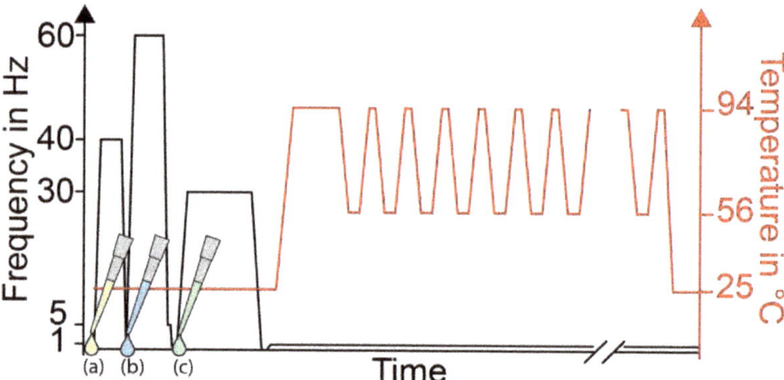

Figure 2. LabDisk frequency and temperature protocol for droplet generation and PCR thermo-cycling in LabDisk Player 1. During droplet generation, the temperature remained constant at room temperature (approximately 25 °C) and the frequency changed, whereas during PCR thermocycling, the frequency remained constant at 1 Hz. For oil transfer and droplet generation, the respective reagents were pipetted into the LabDisk manually, as indicated by a pipet tip (emulsification-oil = yellow (a), reaction mix = blue (b) and PCR-oil = green (c)).

2.4. Automated Fluorescence Imaging and Droplet Evaluation

As shown in Figure 1C, the four-plex ddPCR workflow included the fluorescence imaging of each ddPCR unit with a Lionheart LX (Biotek) microscope. For imaging, each ddPCR unit was scanned individually on a British standard microscope slide (ThermoFisher Scientific). Therefore, the LabDisk was cut into 12 parts, each containing a complete ddPCR unit. Each part was attached to a glass slide with adhesive tape. Although it was possible to scan a complete LabDisk with this microscope, scanning time was reduced with this approach. A standard tray (ID 1450527, Biotek) was used to mount the microscope slides with the ddPCR units onto the x/y movable stage of the microscope. Fifty-six individual images were taken of each reaction chamber and in each fluorescence channel with a 2.5×-magnification lens. The parameters in the imaging software Gen5 (Biotek) were defined so as to stitch and compress the final image to a size of 50%. The imaging settings for each fluorescence channel are listed in Table 1. The fluorescence channels were defined by a standard fluorophore that could be detected in the respective channel. In this work, the wild type DNA target was detected in the green fluorescent pigment (GFP) channel by a FAM fluorophore, the point mutation for *KRAS* G12V in the red fluorescent pigment (RFP) channel by a HEX fluorophore, *KRAS* G12A in the TEXAS RED channel by an AttoRho 101 fluorophore and the *KRAS* G12D sequence in the Cy5 channel by a Cy5 fluorophore.

Table 1. Settings for imaging with a Lionheart LX microscope.

Channel	Led Intensity	Excitation in MS	Gain	Used Fluorophore in ddPCR	Target
GFP	10	300	24	FAM	Wild type
RFP	10	1000	30	HEX	*KRAS* G12V
TEXAS RED	10	500	30	AttoRho 101	*KRAS* G12A
CY5	10	500	30	Cy5	*KRAS* G12D

The same software as for the imaging process was used for droplet evaluation (Gen5, Biotek). The Gen5 software droplet evaluation pipeline is depicted in the ESI. The intended use of the software and the analysis tool is the counting of cells and cell organelles. However, with small adjustments, this tool can also be used to count droplets and to distinguish between positive and negative droplets. For this purpose, the cellular analysis tool was used to define the background, droplet size and droplet shape (circularity > 0.6), as well as thresholds for the fluorescence intensity measurement. It was important to include circularity as a parameter in the calculation metrics so that dust particles or other artifacts, which did not match the defined droplet shape, could be removed. Further inclusion criteria for a detected droplet were the minimum and maximum object sizes (here, 45 μm and 90 μm). To facilitate the target evaluation, each fluorescence channel was defined as a subpopulation. In this work, there were four different subpopulations, as four channels were used. Each subpopulation needed at least two parameters, the circularity and the mean fluorescence intensity, which should be used to differentiate between positive and negative droplets. The line profile tool provided appropriate thresholds for this purpose. A line was drawn across an area of interest, and the intensity profile of each droplet or structure was shown. A fifth subpopulation counted the droplets with a circularity > 0.6 for the overall droplet count. In this work, the GFP channel had the highest background signal in the droplets, so this signal was used to detect the droplets and determine their total number. As we observed the multi-droplet layers in the outer area of the reaction chamber (Figure S4 in ESI), we limited the evaluation to an inner area where droplet monolayers were guaranteed. On average, 6000 droplets were evaluated for each data point during ddPCR validation. These multi-droplet layers could be reduced by redesigning the reaction chamber and, for example, adjusting the depth to a maximum of 140 μm.

3. Results and Discussion

3.1. LabDisk

The designed LabDisk included 12 identical ddPCR units, as shown in Figure 3A. These units could be used either to test different assay conditions, such as primer and probe concentrations, or to test different assay targets. The LabDisk presented here was a single-use cartridge, which was disposed after each run. The target panel of the proposed LabDisk was not fixed and could easily be adapted for different applications. This is a particular advantage of the employed Mediator Probe PCR assay [24]. A limiting factor might be the application-specific requirements regarding droplet number and droplet size. Figure 3B shows a bright field microscopy image of the generated droplets. Robust droplet generation was achieved with the PerfeCTa MultiPlex qPCR Toughmix.

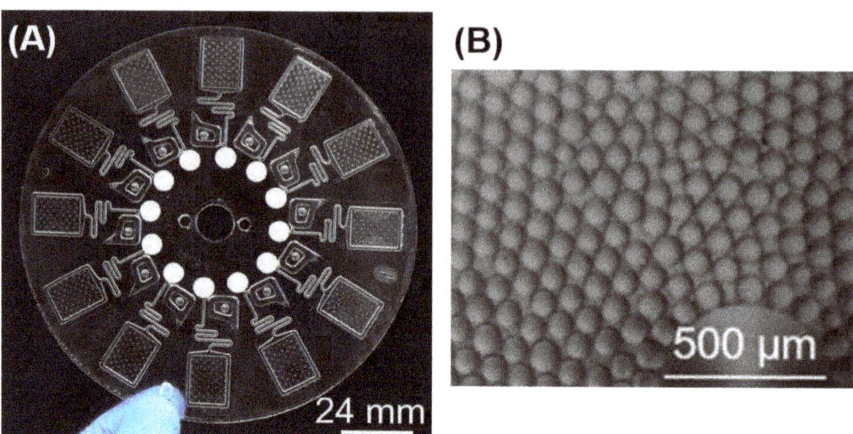

Figure 3. (**A**) The LabDisk with 12 identical ddPCR units manufactured with cyclo-olefin copolymer (COC) foil technology. On the far side, each vent is closed by a white polytetrafluoroethylene (PTFE) membrane. (**B**) Bright field microscopy image of droplets generated with PerfeCTa MultiPlex qPCR Toughmix in a ddPCR unit.

3.2. Evaluation of Surfactant Concentrations in the Emulsification Oil and in the PCR Oil

The 12 identical ddPCR units allowed not only a high number of reactions in a diagnostic application but could also be used to efficiently test and optimize different droplet stabilization settings. The recommended surfactant concentration from the supplier (Fluigent) was 2% for droplet generation. We found this given surfactant concentration quite useful for generating homogeneous droplets. However, during PCR cycling, the oil evaporated completely, and substantial droplet merging occurred (Figure 4A). The vapor pressure of the emulsification-oil was 37,627 Pa at 95 °C [26]. Therefore, after droplet generation, a second oil, termed the PCR-oil, was required for the subsequent cycling process. This second oil must have a lower vapor pressure during cycling, so an oil with a vapor pressure of 9503 Pa at 95 °C was chosen as the PCR-oil [27]. The generated droplets were surrounded by the oil phase, and as the evaporation is an effect of the interface to the surrounding air, the evaporation effect of the oils at the interface was significantly stronger than the evaporation of the droplets. For each ddPCR unit, 6 μL of emulsification-oil and 8 μL of PCR-oil were used. Figure 4A shows the combination of the pure PCR-oil (no surfactant) and different surfactant concentrations in the emulsification-oil. As a general outcome, a high surfactant concentration in the emulsification-oil only (i.e., without surfactant in the PCR-oil) did not prevent droplet merging. In this experiment, surfactant concentrations of 2% to 5% in the emulsification-oil were tested. Figure 4A depicts an exemplary image showing a merged droplet area. As the droplets were generated from a PCR mix which included fluorescence probes and target molecules, the merged droplets appeared brighter than the oil and the background of the image. Covering the droplets with the second oil (PCR-oil) had the effect of the reaction chamber remaining filled with oil after PCR cycling. However, the desired droplet stability was still not achieved. This led to the conclusion that the PCR-oil also needed surfactants to stabilize the droplets in the oil phase during cycling, because evaporation means less of the emulsification-oil remains after every cycle. Therefore, 2% surfactant was dissolved in the PCR-oil, and the experiment with different surfactant concentrations in the emulsification-oil (2–10%) was repeated, as shown in Figure 4B. Two combinations showed no merging (4% and 6% surfactant in the emulsification-oil). The exemplary image in Figure 4B shows a reaction chamber with 4% surfactant in the emulsification-oil and 2% surfactant in the PCR-oil. The emulsification-oil (liquid density at 25 °C, 1.61 kg/L) and the PCR-oil (liquid density at 25 °C, 1.86 kg/L) were miscible, which led to the assumption that the final surfactant concentration in the ddPCR chamber could be averaged. Therefore, it could be worth investigating the surfactant concentration in the emulsification-oil (5–10%) in combination with the pure PCR-oil in further experiments. The resulting surfactant concentration could be strongly dependent on the PCR cycling parameters (temperature and hold times), as the evaporation of the oil is dependent on these parameters. Therefore, the surfactant concentrations in the oil mix should be investigated for each PCR protocol. In this work, the presented combination (4% surfactant in emulsification-oil and 2% surfactant in PCR-oil) was found to be reliable for both droplet generation and droplet stabilization during PCR. A higher surfactant concentration in the emulsification-oil led to slight merging effects, which may have resulted from a partial destabilization of the steric stabilization layer that was generated by the adsorption of surfactant molecules to each droplet [28].

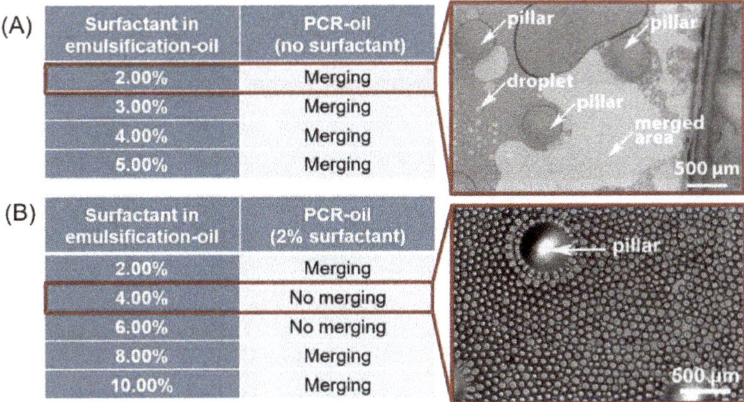

Figure 4. Evaluation of surfactant concentration in emulsification-oil and PCR-oil. (**A**) A mix of 2–5% surfactant in emulsification-oil combined with pure PCR-oil. The microscopy image shows merged areas as bright. (**B**) A mix of 2–10% surfactant in emulsification-oil combined with 2% surfactant in PCR-oil. Exemplary microscopy image shows a droplet area with 4% surfactant in emulsification-oil.

3.3. Demonstration of Four-Plex ddPCR in the LabDisk

The data from Figure 4B show two suitable surfactant concentrations (4% and 6% surfactant in emulsification-oil). As the reagent costs increase with the surfactant concentration, the lower concentration was chosen (4% surfactant in the emulsification-oil and 2% surfactant in the PCR-oil). Lower concentrations (e.g., 3% surfactant) might also work, but were not tested. The aim of this experiment was to determine the performance of a four-plex ddPCR assay inside the LabDisk. Figure 5A shows four-color fluorescence images of an assay with a concentration of 3540 DNA copies of each mutant sequence (*KRAS* G12V (blue), G12A (pink) and G12D (red)) in a background of 15,000 wild type DNA copies (green channel). The proof-of-concept experiment with serial dilutions of the DNA template confirmed efficient PCR amplification of each target at concentrations from 3540 mutant copies down to 35 mutant copies (see Figure 5B; full data in ESI). For the mutation sequences *KRAS* G12V and G12A, even the lowest concentrations of 3.5 mutant copies were detectable (limit of detection). At this low concentration, subsampling errors and the quantification of up to 6000 droplets only, instead of all generated droplets, could be reasons for the missing G12D signal. The discrepancy between the detected copy numbers and the spiked copy numbers per µL could be explained by aliquoting errors and by the fact that assay performance is dependent on the PCR protocol. This could result in sub-optimal annealing temperatures for the different targets, as the annealing temperature was fixed at 56 °C. We observed no contamination of the player or the LabDisk, as shown by negative no-template controls (see ESI). This indicates that the PTFE membrane to close the venting hole was functional. In the future, a mechanically closed chip could additionally reduce any remaining risk of contamination. After imaging and droplet detection, the target concentration (see ESI) was calculated using the Poisson equation [5], and the determination coefficient was calculated using the fitting function in Origin (OriginLab). For the calculation of the target concentration, the droplet diameter has to be evaluated. During droplet detection with the Gen5 software, the diameter of each droplet was measured automatically in the fluorescence image. Compared with a bright field image, the fluorescence image provides much better contour detection, because the droplets appear bright on a dark background. The droplet diameter of 82.7 ± 9 µm (histogram; see ESI) was determined with the Lionheart LX microscope (n = 6000 droplets). The 10 µL of PCR mix led to a final number of approximately 35,000 droplets for each ddPCR unit, as calculated from a measured droplet size of about 83 µm. Each mutation assay showed very high reaction linearity ($R^2 = 0.99$) within the determined limit of quantification (LoQ = 35 copies per reaction). This proves the functionality of the four-plex ddPCR in the LabDisk and demonstrates its potential for application in cancer monitoring. The readout of all generated droplets could decrease the subsampling error and thus improve the LoQ.

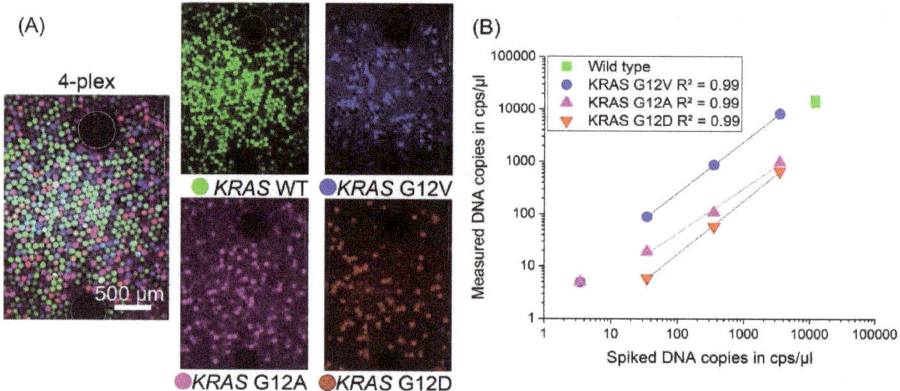

Figure 5. (**A**) Fluorescence image overlay of the four-plex ddPCR assay for the *KRAS* wild type (WT) and the point mutations *KRAS* G12D, G12V and G12A. Individual fluorescence images show amplification signals in all three color channels for the mutant DNA and in the fourth color channel for the wild type DNA. (**B**) The proof-of-concept serial dilution experiment showed the high reaction linearity of the four-plex ddPCR assay inside the LabDisk across 3540–35 mutant DNA copies per µL. The lowest dilution point (3.5 mutant DNA copies per µL) was quantifiable for *KRAS* G12V and G12A, but not for G12D.

4. Conclusions and Outlook

This work emphasizes the relevance of surfactant concentrations and oil combinations for achieving stable droplets during droplet generation and PCR thermocycling. Moreover, the presented image evaluation pipeline shows the simplicity of using a commercial fluorescence microscope for droplet counting and four-color ddPCR readout. The introduced four-plex ddPCR workflow on a LabDisk offers great potential for further ddPCR applications. Although shown for cancer targets here, the proposed LabDisk could be used for a broad range of ddPCR applications focusing on the quantification of very low concentrations of point mutations or other DNA sequences in a high wild type DNA background. With these results, we lay the platform for the realization of digital, centrifugal microfluidic sample-to-answer systems for multiplex applications. We think a combination of the presented ddPCR unit with a microfluidic DNA extraction unit on the same cartridge is of particular interest for PoC testing scenarios.

Supplementary Materials: The following are available online at https://www.mdpi.com/2227-9717/9/1/97/s1. Preparation of gBlocks. Instruction for HaeIII digestion of human genomic DNA. Supplemental Figure S1. Image evaluation pipeline for droplet counting with Lionheart software Gen5. With the cellular analysis tool, the droplet counting in each ddPCR unit is simplified, as it is an automated process. Important parameters for defining the droplet contour are background values, droplet diameter and droplet shape (circularity). Each fluorescence channel defines a DNA target sequence, and thus it is necessary to adjust the threshold for each channel to determine the positive and negative droplet count. Supplemental Table S1: ddPCR data for four-plex assay. Supplemental Figure S2: images of ddPCR areas after thermocycling. Each picture (A–D) shows a dilution step of synthetic mutant gBlocks in human genomic DNA. Supplemental Figure S3: negative template control of the four-plex ddPCR assay. Image overlay of four-color fluorescence images. Bright artifacts are caused by dust particles on the cartridge surface. Supplemental Figure S4: droplet evaluation in the ddPCR unit. Yellow circles indicate droplets detected by the Gen5 software. The monolayer in the center of the ddPCR area is used for droplet counting. Bright regions in the outer parts of the ddPCR area are due to droplet overlay (multi-droplet layers). Supplemental Figure S5: (A) PMMA cover for thermocycling with cut-outs between each ddPCR unit, and (B) PMMA cover on top of LabDisk to shield ddPCR units from potential temperature overshoots caused by peaks in the dynamically controlled LabDisk Player 1 air heating module's output. Supplemental Figure S6: droplet diameter histogram for the evaluation of 6000 droplets in the green fluorescence channel. The monolayer in the chamber is used for droplet counting and threshold detection.

Author Contributions: Conceptualization, F.S., P.J. and T.H.; methodology, F.S., P.J. and T.H.; Mediator Probe ddPCR assay design and optimization, E.K.; investigation of ddPCR on LabDisk, F.S.; writing—original draft preparation, F.S.; writing—review and editing, all authors; supervision, R.Z., F.v.S., N.P., N.B., P.J. and T.H.; project administration, P.J.; funding acquisition, T.H. All authors have read and agreed to the published version of the manuscript.

Funding: Funding from the Federal Ministry of Education and Research (BMBF), Germany, within the project Artur (grant number 13GW0198F; project management organization VDI Technologiezentrum GmbH), and from the Ministry of Economic Affairs, Labor and Housing of the State of Baden-Württemberg, Germany, within the project PRIMO, is gratefully acknowledged.

Institutional Review Board Statement: Not applicable.

Informed Consent Statement: Not applicable.

Data Availability Statement: Not applicable.

Acknowledgments: We thank the team of the Hahn-Schickard foundry service for the fabrication of the LabDisks. In addition, we want to thank the team of the liquid biopsy lab and the core facility at the University Hospital Freiburg, especially Ulrike Philipp, Max Deuter, Marie Follo, Saskia Hussung and Nikolas von Bubnoff, for supporting the ddPCR assay design and providing the target sequences.

Conflicts of Interest: The authors declare no conflict of interest. The funders had no role in the design of the study; in the collection, analysis or interpretation of the data; in the writing of the manuscript; or in the decision to publish the results.

References

1. Perkins, G.; Lu, H.; Garlan, F.; Taly, V. Droplet-Based Digital PCR: Application in Cancer Research. *Adv. Clin. Chem.* **2017**, *79*, 43–91. [CrossRef] [PubMed]
2. Sreejith, K.R.; Ooi, C.H.; Jin, J.; Dao, D.V.; Nguyen, N.-T. Digital polymerase chain reaction technology—Recent advances and future perspectives. *Lab Chip* **2018**, *18*, 3717–3732. [CrossRef] [PubMed]
3. Olmedillas-López, S.; García-Arranz, M.; García-Olmo, D. Current and Emerging Applications of Droplet Digital PCR in Oncology. *Mol. Diagn. Ther.* **2017**, *21*, 493–510. [CrossRef] [PubMed]
4. Baker, M. Digital PCR hits its stride. *Nat. Methods* **2012**, *9*, 541–544. [CrossRef]
5. Basu, A.S. Digital Assays Part I: Partitioning Statistics and Digital PCR. *SLAS Technol.* **2017**, *22*, 369–386. [CrossRef] [PubMed]
6. Sundberg, S.O.; Wittwer, C.T.; Gao, C.; Gale, B.K. Spinning disk platform for microfluidic digital polymerase chain reaction. *Anal. Chem.* **2010**, *82*, 1546–1550. [CrossRef]
7. Li, Z.; Liu, Y.; Wei, Q.; Liu, Y.; Liu, W.; Zhang, X.; Yu, Y. Picoliter Well Array Chip-Based Digital Recombinase Polymerase Amplification for Absolute Quantification of Nucleic Acids. *PLoS ONE* **2016**, *11*, e0153359. [CrossRef]

8. FORTUNE. Market Research Report. Digital PCR Market Size, Share & Industry Analysis, By Type (Droplet Digital PCR, Chip-based Digital PCR, and Others), By Product (Instruments and Reagents & Consumables), by Indication (Infectious Diseases, Oncology, Genetic Disorders, and Others), by End User (Hospitals & Clinics, Pharmaceutical & Biotechnology Industries, Diagnostic Centers, and Aca-demic & Research Organizations), and Regional Forecast 2019–2026. Digital PCR Market FBI102014. 2020. Available online: https://www.fortunebusinessinsights.com/digital-pcr-market-102014 (accessed on 9 November 2020).
9. Schuler, F.; Trotter, M.; Geltman, M.; Schwemmer, F.; Wadle, S.; Domínguez-Garrido, E.; López, M.; Cervera-Acedo, C.; Santibáñez, P.; von Stetten, F.; et al. Digital droplet PCR on disk. *Lab Chip* **2016**, *16*, 208–216. [CrossRef]
10. Hindson, B.J.; Ness, K.D.; Masquelier, D.A.; Belgrader, P.; Heredia, N.J.; Makarewicz, A.J.; Bright, I.J.; Lucero, M.Y.; Hiddessen, A.L.; Legler, T.C.; et al. High-throughput droplet digital PCR system for absolute quantitation of DNA copy number. *Anal. Chem.* **2011**, *83*, 8604–8610. [CrossRef]
11. Hiddessen, A.L.; Chia, E.R. Stabilized Droplets for Calibration and Testing. U.S. Patent 10,240,187, 26 March 2019.
12. Madic, J.; Zocevic, A.; Senlis, V.; Fradet, E.; Andre, B.; Muller, S.; Dangla, R.; Droniou, M.E. Three-color crystal digital PCR. *Biomol. Detect. Quantif.* **2016**, *10*, 34–46. [CrossRef]
13. Strohmeier, O.; Keller, M.; Schwemmer, F.; Zehnle, S.; Mark, D.; von Stetten, F.; Zengerle, R.; Paust, N. Centrifugal microfluidic platforms: Advanced unit operations and applications. *Chem. Soc. Rev.* **2015**, *44*, 6187–6229. [CrossRef] [PubMed]
14. Burger, R.; Amato, L.; Boisen, A. Detection methods for centrifugal microfluidic platforms. *Biosens. Bioelectron.* **2016**, *76*, 54–67. [CrossRef] [PubMed]
15. Geissler, M.; Brassard, D.; Clime, L.; Pilar, A.V.C.; Malic, L.; Daoud, J.; Barrère, V.; Luebbert, C.; Blais, B.W.; Corneau, N.; et al. Centrifugal microfluidic lab-on-a-chip system with automated sample lysis, DNA amplification and microarray hybridization for identification of enterohemorrhagic Escherichia coli culture isolates. *Analyst* **2020**, *145*, 6831–6845. [CrossRef] [PubMed]
16. Chen, Z.; Liao, P.; Zhang, F.; Jiang, M.; Zhu, Y.; Huang, Y. Centrifugal micro-channel array droplet generation for highly parallel digital PCR. *Lab Chip* **2017**, *17*, 235–240. [CrossRef] [PubMed]
17. Schulz, M.; Probst, S.; Calabrese, S.; Homann, A.R.; Borst, N.; Weiss, M.; von Stetten, F.; Zengerle, R.; Paust, N. Versatile Tool for Droplet Generation in Standard Reaction Tubes by Centrifugal Step Emulsification. *Molecules* **2020**, *25*, 1914. [CrossRef] [PubMed]
18. Schulz, M.; Calabrese, S.; Hausladen, F.; Wurm, H.; Drossart, D.; Stock, K.; Sobieraj, A.M.; Eichenseher, F.; Loessner, M.J.; Schmelcher, M.; et al. Point-of-care testing system for digital single cell detection of MRSA directly from nasal swabs. *Lab Chip* **2020**, *20*, 2549–2561. [CrossRef]
19. Li, X.; Zhang, D.; Ruan, W.; Liu, W.; Yin, K.; Tian, T.; Bi, Y.; Ruan, Q.; Zhao, Y.; Zhu, Z.; et al. Centrifugal-Driven Droplet Generation Method with Minimal Waste for Single-Cell Whole Genome Amplification. *Anal. Chem.* **2019**, *91*, 13611–13619. [CrossRef]
20. Liao, P.; Jiang, M.; Chen, Z.; Zhang, F.; Sun, Y.; Nie, J.; Du, M.; Wang, J.; Fei, P.; Huang, Y. Lossless and Contamination-Free Digital PCR. *bioRxiv* **2019**, 739243. [CrossRef]
21. Focke, M.; Stumpf, F.; Faltin, B.; Reith, P.; Bamarni, D.; Wadle, S.; Müller, C.; Reinecke, H.; Schrenzel, J.; Francois, P.; et al. Microstructuring of polymer films for sensitive genotyping by real-time PCR on a centrifugal microfluidic platform. *Lab Chip* **2010**, *10*, 2519–2526. [CrossRef]
22. POREX. PTFE Protection Vents. PMV15N. 2015. Available online: https://www.porex.com/wp-content/uploads/2020/01/Porex-Datasheet-Virtek-IP-Rated-Protection-Vents.pdf (accessed on 20 November 2020).
23. Faltin, B.; Wadle, S.; Roth, G.; Zengerle, R.; Stetten, F. von. Mediator probe PCR: A novel approach for detection of real-time PCR based on label-free primary probes and standardized secondary universal fluorogenic reporters. *Clin. Chem.* **2012**, *58*, 1546–1556. [CrossRef]
24. Lehnert, M.; Kipf, E.; Schlenker, F.; Borst, N.; Zengerle, R.; Stetten, F. von. Fluorescence signal-to-noise optimisation for real-time PCR using universal reporter oligonucleotides. *Anal. Methods* **2018**, *10*, 3444–3454. [CrossRef]
25. Fluigent. dSURF Product Data Sheet. Available online: https://www.fluigent.com/wp-content/uploads/2018/10/SCL_OPE_001_A-Data-sheet-dSURF.pdf (accessed on 20 November 2020).
26. 3M. 3M Novec 7500 High-Tech Flüssigkeit. 2014. Available online: https://multimedia.3m.com/mws/media/1438121O/3m-novec-7500-engineered-fluid-tech-sheet.pdf (accessed on 20 November 2020).
27. 3M. Fluorinert. Electronic Liquid FC-40. Product Information. 2019. Available online: https://multimedia.3m.com/mws/media/64888O/3m-fluorinert-electronic-liquid-fc40.pdf (accessed on 20 November 2020).
28. Pichot, R. Stability and Characterisation of Emulsions in the Presence of Colloidal Particles and Surfactants. Ph.D. Thesis, University of Birmingham, Birmingham, AL, USA, 2012.

Article

Characterization of a Wireless Vacuum Sensor Prototype Based on the SAW-Pirani Principle

Sofia Toto †, **Mazin Jouda** †, **Jan G. Korvink** †, **Suparna Sundarayyan**, **Achim Voigt** †, **Hossein Davoodi** † and **Juergen J. Brandner** *,†

Institute of Microstructure Technology, Karlsruhe Institute of Technology, 76131 Karlsruhe, Germany; sofiatoto5@hotmail.com (S.T.); mazin.jouda@kit.edu (M.J.); jan.korvink@kit.edu (J.G.K.); susu1012@hs-karlsruhe.de (S.S.); achim.voigt@kit.edu (A.V.); hossein.davoodi@kit.edu (H.D.)

* Correspondence: juergen.brandner@kit.edu; Tel.: +49-721-608-23963

† Current address: Hermann von Helmholtz Platz 1, 76344 Eggenstein Leopoldshafen, Germany.

Received: 31 October 2020; Accepted: 3 December 2020; Published: 21 December 2020

Abstract: A prototype of a wireless vacuum microsensor combining the Pirani principle and surface acoustic waves (SAW) with extended range and sensitivity was designed, modelled, manufactured and characterised under different conditions. The main components of the prototype are a sensing SAW chip, a heating coil and an interrogation antenna. All the components were assembled on a 15 mm × 11 mm × 3 mm printed circuit board (PCB). The behaviour of the PCB was characterised under ambient conditions and in vacuum. The quality of the SAW interrogation signal, the frequency shift and the received current of the coil were measured for different configurations. Pressures between 0.9 and 100,000 Pa were detected with sensitivities between 2.8 GHz/Pa at 0.9 Pa and 1 Hz/Pa close to atmospheric pressure. This experiment allowed us to determine the optimal operating conditions of the sensor and the integration conditions inside a vacuum chamber in addition to obtaining a pressure-dependent signal.

Keywords: SAW; Pirani; compact; wireless; vacuum; sensing

1. Introduction

Many industry and research facilities have some of their processes performed under vacuum conditions. Those include semiconductor technology plants, food industry sites and the aerospace industry sites. The existence of vacuum installations induces the need for reliable vacuum pressure monitoring.

A pressure value can provide valuable information about a process and by means of the detection of some gaseous species or leakage, which enables process diagnosis.

Miniaturization goes hand in hand with microfluidics. The designed system contains of a cube crossed by a microchannel where the sensing chip is located. The heat transfer between the chip and its surrounding microchannel allows one to perform the sensing operation.

The miniaturization of sensors relies on microfluidics and allows low cost, low power and non-invasive sensing. In the case of Pirani sensing, the measuring is enabled by device miniaturization. Indeed, the signal to noise ratio is increased by reducing solid thermal conduction.

A vacuum refers to any condition at subatmospheric pressure, which can range from a rough vacuum close to atmospheric pressure to an extremely strong vacuum around 10^{-10} Pa. Handling such a wide range usually involves several pressure transducers simultaneously integrated inside a chamber. Multiple sensor use in a vacuum chamber raises many issues: integration, maintenance, redundancy, power consumption, wiring and readout to name a few. Extending the sensing range using a single miniature device operating wirelessly is therefore a viable alternative.

The Pirani principle is commonly used to sense pressure in the fine and rough vacuum range. It is based on the heat transfer between a heated sensing element (wire, plate or chip) and its surrounding gas molecules. Since the heat transfer is proportional to the number of molecules, the temperature variation of the sensor depends on the pressure. Heating is necessary to observe a pressure induced temperature variation.

Surface acoustic waves (SAW) propagate on the surfaces of piezoelectric crystals. An interdigital transducer (IDT), which is a set of metallic electrodes etched on the surface of the piezoelectric substrate, converts electrical voltage into waves back and forth. SAW are sensitive to the environment properties like temperature, pressure and humidity. When a material is heated, it is subject to thermal expansion. Its physical properties such as its elasticity, piezoelectricity and permittivity are modified. The variations of those properties are mostly expressed as Taylor series expansions. The resonance frequency of a SAW IDT can be expressed as:

$$f = \frac{V}{\lambda} \tag{1}$$

where V is the wave propagation velocity and λ the wavelength which depends on the electrode's dimensions. At the first order, the sensitivity of the resonance frequency f_r to the temperature around a reference temperature T_0 is expressed by the coefficient TC_{f1} (in ppm/K):

$$\frac{\Delta f}{f_r\,(T_0)} = TC_{f1}\,(T - T_0) \tag{2}$$

where T is the temperature and Δf is the resonance frequency shift. In [1], a high-sensitivity wireless temperature SAW sensor is presented. Besides, SAW devices can be easily wirelessly operated [2].

Several wireless SAW-Pirani sensors are presented in the literature, such as [3–5]. In [6], a GaN membrane SAW sensor that measures simultaneously the pressure and the temperature with the same SAW structure operating as a dual sensor is demonstrated. Sensitivities higher than 60 MHz/MPa for the Lamb mode and about 20 MHz/MPa for the Rayleigh mode were obtained for the structure having fingers and pitches which were 170 nm wide. These values correspond to pressure coefficients of frequency (PCF) of about 3500 ppm/MPa for the Rayleigh and 5400 ppm/MPa for the Lamb mode values, exceeding the values previously reported for SAW pressure sensors. SAW sensors as transducers are shown to be suitable for measuring pressures higher than 1 atm.

In [7], a SAW pressure sensor is presented. The evaluated sensitivities were 87.81 ppm/K, 900 ppm/MPa and 0.0023 dB/Ω.

With this in mind, a miniaturised wireless vacuum sensor with extended range and sensitivity was designed, simulated and prototyped [8,9]. The assembly of a prototype entailed challenges linked to the operation of two different electromagnetic waves, the miniaturization of all the components and wireless power transfer in a vacuum environment while most components were made of stainless steel. These challenges raised the question of the feasibility of such a solution and its convenience. After manufacturing, assembly issues were addressed and the first characterisation tests were performed.

2. Materials and Methods

The objective of the measurements performed in vacuum was to be able to observe a pressure-dependent signal, i.e., measure a resonance frequency shift vs. pressure through all the vacuum pump pressure range. After characterising the prototype at room temperature and pressure, the prototype was tested and characterised inside a vacuum chamber.

2.1. Description of the Test Rig:

The vacuum test rig shown in Figure 1 was made of the following components:

- Vacuum pump: a TSH071E Turbocube from Pfeiffer Vacuum (Aßlar, Germany) was used to generate subatmospheric pressure conditions.
- Metallic ISO KF pipes (METALLIC FLEX, Kirchweg, Germany) were used to connect the components.
- A VAT gate valve was used to protect the turbo pump in the rough vacuum region between 10 Pa and atmospheric pressure since it cannot operate under a pressure higher than 10 Pa.
- A BCG 450 triple gauge from INFICON (Bad Ragaz, Switzerland) was used as a pressure sensor to monitor the pressure: it contained a Bayard Alpert gauge, a Pirani gauge and a capacitive diaphragm gauge with automatic switch to measure pressures between 5×10^{-8} Pa and 1.5×10^{5} Pa. The pressure sensor contained a display indicating the real time pressure and was connected to a LabVIEW user interface via an RS232 to USB cable.
- A needle valve was used to control the pressure.
- A window box was prepared to operate the prototype. The sensor was placed inside the box against the window from inside, whereas the interrogation antenna and Tx coil were placed against the window from outside. The PMMA window enabled the propagation of the waves towards and from the vacuum.

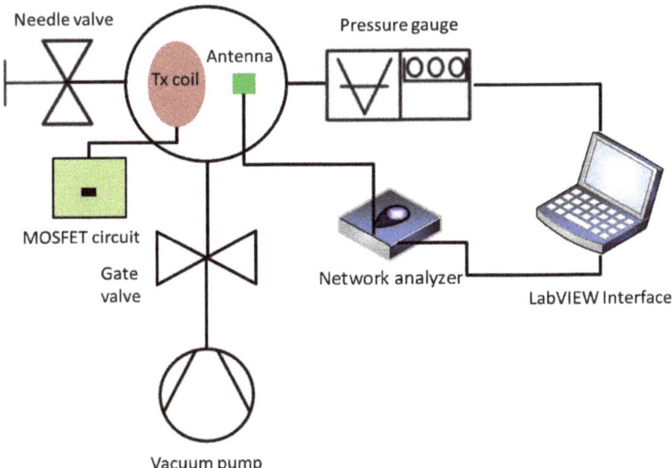

Figure 1. Schematic of the vacuum test rig showing its components.

In order to test the prototype in vacuum, adjustments to the vacuum chamber were required. It mainly consisted in the preparation of a vacuum PMMA window. Due to the fact that the vacuum viewports available off the shelf are made of borosilicate with a metallisation layer, they are not suitable for SAW interrogation. Companies like VACOM or INFICON commercialise viewports for vacuum optics of several materials and dimensions. The standard viewports are made of borosilicate and fused silicate and show transmission spectra for wavelengths between 0.2 µm and 5 µm. The 2.45 GHz SAW wavelength is 1.3836 µm. The wavelength of the induction wave for coils coupling is in the kilometer range. At first glance the viewport should allow the propagation of the SAW and not the induction wave. However, to prevent tensions occurring during heating, cooling or installing, the viewport contains an intermediate metallisation layer of Kovar (an iron-nickel-cobalt alloy) that prevents electromagnetic waves propagation.

For this reason, a customised polymethyl methacrylate (PMMA) window was manufactured to test the prototype in vacuum. The PMMA circular window had an outer diameter of 160 mm corresponding to the previous viewport diameter and the chamber diameter. The central part of the window, i.e., its 5 cm diameter central circle, had a 2 mm thickness. The transmitting (Tx) coil and Tx antenna were placed outside facing the prototype and the receiving (Rx) coil. Reducing the thickness to 2 mm was done to increase the quality of the signal delivered by the antenna. Since the structure of the chamber did not allow one to put the transmitting coil and the interrogation antenna on different sides, the heating power transfer was separated from the antenna interrogation. The receiver coil was connected to a 10 Ω surface mounted device (SMD) resistor which was soldered on top of the chip.

The prototype inserted inside vacuum consisted mainly of a PCB. On top of the PCB the receiver coil and antenna were mounted. At the bottom of the PCB, the SAW chip was placed. The SMD resistor soldered on top of the SAW chip and connected to the receiver coil was acting as a Joule heating resistance. The soldering of the resistor provided a good thermal contact between the resistor and the SAW chip. A photo of the setup is shown in Figure 2.

Since the structure of the chamber did not allow placing the transmitting coil and the interrogation antenna at different sides, as could be possible using a smaller ceramic or glass box with an ISO KF or CF junction, the receiver coil was separated from the rest of the PCB. The heating was performed via an SMD resistor to substitute the direct heating of the chip via a gold metallic layer joule resistance at its bottom. The temperature was measured by placing a Pt1000 sensor in the vicinity of the chip inside the vacuum.

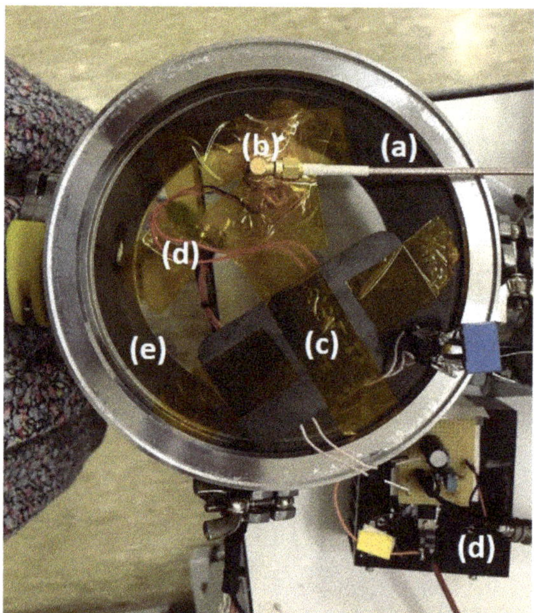

Figure 2. Top view of the test rig showing the window. The Tx coil and the Tx antenna are placed on top of the window. (**a**) SMA cable going to the network analyser. (**b**) Sensor PCB. (**c**) Transmitting coil. (**d**) Cable inside vacuum connecting the receiver coil and the heating resistor. (**e**) PMMA window.

2.2. *Experimental Procedure*

The objective of the test rig was to measure the resonance frequency of the SAW chip versus pressure. After the assembly of the test rig the vacuum pump was turned on. When the ultimate pressure was reached, the pressure was gradually increased by turning the needle valve. The corresponding resonance frequency of the chip was recorded. Figure 3 shows the complete setup.

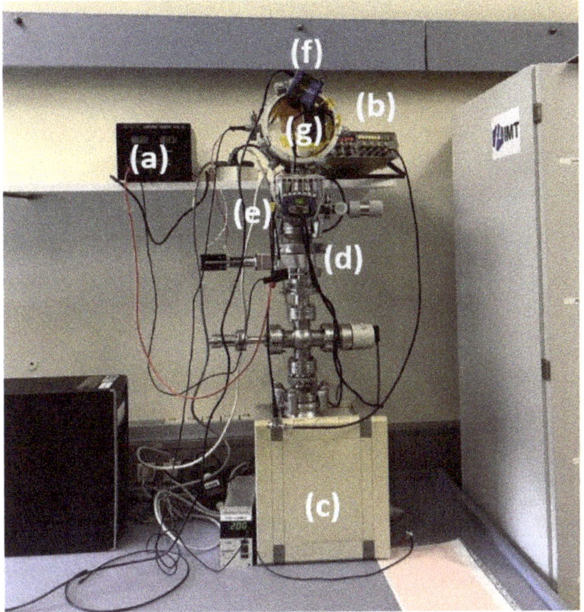

Figure 3. Complete test rig: (**a**) Voltage source. (**b**) Arbitrary waveform generator. (**c**) Vacuum pump. (**d**) Valve to protect the turbo pump. (**e**) Pressure sensor. (**f**) Vector network analyser. (**g**) Window setup shown in Figure 2.

The pressure was varied between 0.9 and 100,000 Pa and the behaviour of the chip was observed during one hour by means of a vector network analyser (VNA). Each pressure was maintained over the course of one hour and VNA interrogations were performed every 5 min in order to obtain 10 measurement points for each pressure. Two points per pressure decade were measured. For each pressure value, the measurements were performed when the pressure was stabilised, i.e., 5 min after turning the needle valve to modify the pressure.

The stabilisation of the SAW peak frequency was tracked. The SAW peak was identified from the reflection coefficient S_{11} vs. frequency curve via a gain local minimum and a phase local maximum. The MOSFET of the heating unit Tx circuit outside the chamber was supplied with a 16 V square wave at 157 kHz and 0.84 A. The measured output at the receiver coil and heating resistor was 700 mV RMS and 2 V peak to peak. The power provided to the heating resistance was calculated from the voltage measured, which equals:

$$P = \frac{U^2}{R} = \frac{0.7^2}{10} = 49 \text{ mW} \quad (3)$$

2.3. Calibration

Prior to the chip interrogation, the network analyser had to be calibrated. Through the use of scattering parameters, the device under test (DUT), i.e., the SAW chip, could be characterised by determining its resonance frequency. At high frequency, scatter parameters are directly related to gain, return loss and reflection coefficient. A VNA's performance depends on the accuracy of its measurements, which require a preliminary calibration. The calibration process aims to perform a vector error correction to remove errors from actual measurements. An analogy to a VNA calibration process is the zeroing out of a test-lead resistance from an ohmmeter to remove the offset caused by the lead resistance.

The error model of a VNA includes multiple terms for every frequency at which measurements are made. The frequencies where the calibration measurements are made must be carefully chosen in coordination with the frequency range where the VNA is to be used. The calibration process determines the error terms, requires a VNA, cables, and standard probes with known impedance. The calibration is performed by sequentially making measurements using calibration standards. Reliable and suitable connectors and cables handled with care are critical components of the calibration setup. No stresses exceeding their specifications should be applied on them. These calibration standards are one-port and two-port networks that have known characteristics. The one port network is used for reflection mode and the two port network is used for transmission mode. The characterisation of the SAW chip needed only the reflection coefficient S_{11}, hence a reflection mode with one port network. For this purpose, a three steps calibration standard using a coaxial cable was applied. The three steps are:

- Short (S): A coaxial short has a total reflection of magnitude 1. The reflection coefficient of the short is dependent only on its length offset, which represents the length between the reference plane and the short. The loss occurring over this length can generally be ignored. Modelling the short in a VNA requires that only its electrical length be entered into the instrument, but in some cases the model can be extended using the polynomial coefficients to account for parasitic inductance.
- Open (O): A coaxial open standard is constructed using a closed design to avoid effects caused by entry of stray electromagnetic energy. At the open end of the inner conductor, a frequency-dependent fringing capacitance is formed. Even if an open standard could physically be constructed with a length of 0, fringing capacitances would result in a negative imaginary part for S_{11} at higher frequencies.
- Match (M): A match is a precision broadband of 50 Ω impedance that has a value corresponding to the system impedance.

The open-short-match technique described above is the most popular for one-port calibration. It uses the three standards connected one after the other to the test port and the relations occurring for wave-quantities a1 to b1 are determined. The behaviour of the individual standards is assumed to be known. The corresponding descriptive models are stored in the vector network analyser. Based on the three derived values, the three unknown quantities (source match, directivity, and reflection tracking) can be calculated.

The three calibration probes were included with the VNA. A calibration was performed for every new set of measurements (position changing, modification of the test rig). The SAW peak frequency was always situated between 2.4 and 2.5 GHz. The frequency sweep was performed first between 2 and 3 GHz and then between 2.4 and 2.5 GHz. The calibration was adjusted accordingly; i.e., the number of calibration points was increased from 1000 to 30,000 between 2 and 3 GHz in the calibration file of the VNA. Frequency steps of 1 kHz were generated. A smoothing algorithm was applied to all the measurement points. A smoothing algorithm is an averaging function that works over a single sweep. The percentage, or aperture, is adjustable up to a maximum of approximately 20–25% of the span (depending on the specific network analyser). The result is a smoothing of the trace by the vector network analyser and the outcome is similar to video bandwidth filtering in a spectrum analyser. The smoothing reduces the noise and increases the signal to noise ratio (SNR).

For each frequency point and its associated window the algorithm normally looks for the mid-point using adjacent data. However, at the beginning and end points (of the measurement span), the algorithm performs a look ahead and look behind. This provides the smoothing function the same number of points for the smoothing window at the beginning and end of the sweep. For noise reduction, averaging is more appropriate, allowing noise to be averaged out. The main disadvantage is that increased averaging factors increase the required measurement time. For averaging, VNAs use sweep-to-sweep averaging in which each data point is computed based on an exponential average of consecutive sweeps weighted by a user-specified averaging factor. Each new sweep is averaged into the trace until the total number of sweeps is equal to the averaging factor for a fully averaged trace. Each point on the trace is the vector sum of the current trace data and the data from the previous sweep. A high averaging factor gives the best signal-to-noise ratio but slows the trace update time. Averaging applies to each s-parameter per measurement channel, no matter which s-parameters are actually displayed. Regardless of the sweep type, the results of the averaged/corrected trace were placed in the analyser's internal data arrays.

3. Results

The first characterisation of the prototype in vacuum aimed to evidence a pressure-dependent behaviour. This setup allowed to obtain wirelessly a pressure-dependent measurement signal via the use of a miniaturised coil and antenna in vacuum and validates the sensor feasibility and concept of miniaturised wireless vacuum sensor.

Different windows were tested. The standard viewport from INFICON made of borosilicate with a metallisation layer and 10 mm thickness did not grant any SAW signal. The rest of the measurements were performed using the custom made PMMA window described in the materials and methods section. The operating frequency chosen for the heating unit, 157 kHz, is the one that gave maximum output voltage at the heating resistor. The voltage output at the receiver coil that is directly provided to the 5 Ω heating resistance equals to 0.7 V RMS, which corresponds to 100 mW of power approximately. The simulated chip needs 24 mW. The requirements of the heating unit were fulfilled. No SAW peak frequency shift was observed if there was no heating. The observed response time was 1 min approximately. The SAW peak frequency was stabilised one minute after the pressure was modified. The response times calculated (100 s) are therefore realistic. At some pressures, two SAW peaks instead of one were observed in a repeatable way. The behaviour of the antenna signal itself depends on the vacuum pressure. At some pressures the antenna also shows two resonance peaks instead of only one, as shown in Figure 4. The second peak may correspond to a harmonic of the main resonance frequency or some reflection phenomenon.

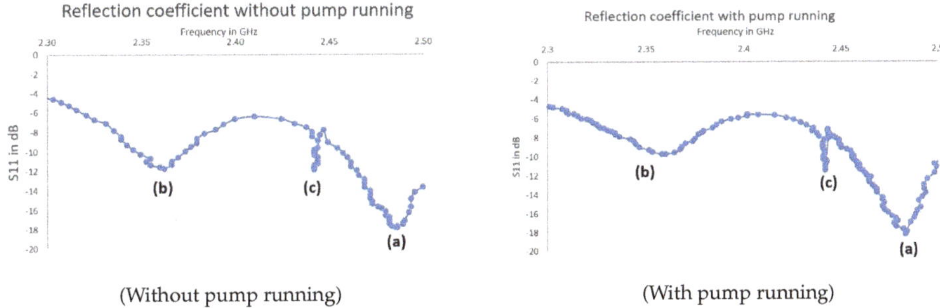

(Without pump running) (With pump running)

Figure 4. Comparison of the antenna signal with a pump running and with no pump running: (**a**) The antenna's self resonance, main peak. (**b**) The antenna's self resonance, second peak. (**c**) Surface acoustic waves (SAW) chip resonance peak. RL is the gain and RP is the phase. Peaks are less sharp with the pump running.

From Figure 4, it can be seen that the operation of the pump influences the signal of the antenna. A secondary self resonance of the antenna can be seen left from the SAW peak. The SAW peak is located between 2 antenna self resonances. A rule of thumb criterion to identify a resonance is an S_{11} value below -15 dB. In the upper graph (a) only the second self resonance of the antenna approaches this value, the SAW peak and the first antenna self resonance approaches -10 dB.

A pressure-dependent signal was obtained. When the pressure increases the temperature of the chip decreases. When the temperature decreases the resonance frequency increases. When the pressure increases the resonance frequency increases as well. The criteria to identify the SAW peak frequency were a local minimum of the gain and a local maximum of the phase. The sensitivity is higher between

0.1 and 1 kPa than close to atmospheric pressure. The results obtained with the prototype are shown in Figure 5.

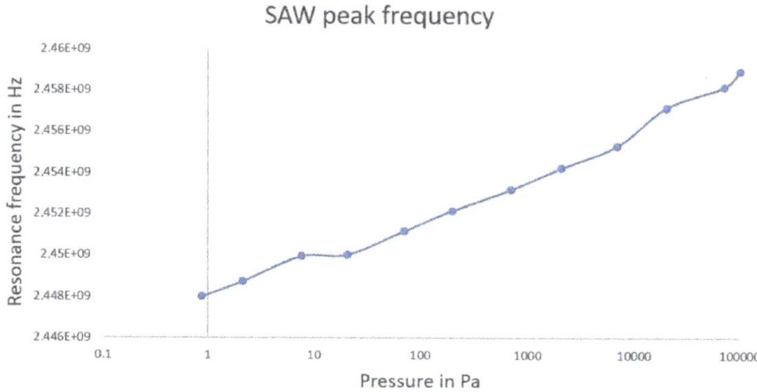

Figure 5. SAW peak frequency vs. pressure.

The sensitivity s_P of the sensor to pressure is defined as:

$$s_P = \left.\frac{df_{res}}{dP}\right|_{T=const} \tag{4}$$

where f_{res} is the resonance frequency of the chip; P the pressure and T the temperature; the pressure coefficient of frequency PCF is:

$$PCF = \left.\frac{1}{f_{res}}\frac{df_{res}}{dP}\right|_{T=const} \tag{5}$$

where f_{res} is the resonance frequency of the chip, P the pressure and T the temperature. The sensors shows a linear behaviour.

The temperature sensitivity s_T is defined as:

$$s_T = \left.\frac{df_{res}}{dT}\right|_{P=const} \tag{6}$$

where f_{res} is the resonance frequency of the chip and T the temperature.

The temperature coefficient of frequency TCF is defined as:

$$TCF = \left.\frac{1}{f_{res}}\frac{df_{res}}{dT}\right|_{P=const} \tag{7}$$

where f_{res} is the resonance frequency of the chip and T the temperature. If the frequency shift vs. pressure and temperature is linear, the resonance frequency can be expressed as:

$$f = f_0 + s_T\,(T - T_0) + s_P\,(P - P_0) \tag{8}$$

where f_0 is the resonance frequency at the temperature T_0 and the pressure P_0; s_T is the temperature sensitivity; and s_P is the pressure sensitivity.

The sensitivity in Hz/Pa changes for the different pressure ranges and is reported in Table 1. The average sensitivity is 5880 Hz/Pa, i.e., 2.4 ppm/Pa between 0.9 Pa and atmospheric pressure. Inside the vacuum, the resonance frequency of the SAW chip measured changed with respect to pressure. The resonance frequency increased with the pressure between 244,7989,448 Hz at 0.871853 Pa and 2,448,780,148 Hz at 99,722.23 Pa. The sensitivity constantly decreased with the pressure according to the expectations between approximately 42 kHz/Pa at 0.9 Pa and 1 Hz/Pa at atmospheric pressure. The sensitivity is higher in the rough vacuum area between the minimum pressure and 200 Pa approximately corresponding to the limit between the Knudsen regime and the transition regime [8], which corresponds of Knudsen number close to 1 [10]. The Knudsen regime is not explained in detail here but in [11]. The uncertainty of the pressure sensor used as a reference is 15% and the uncertainty of the frequency was taken as a first approximation as the double of the standard deviation of the measurements. Since those measurements constitute only a proof of concept, no extended uncertainty analysis was performed. The objective was to observe the rough signal measured in vacuum and determine if the behaviour of the chip is influenced by the value of the vacuum pressure. Table 1 and Figure 6 show the computed sensitivity of the sensor.

Table 1. Computed sensitivity of the resonance frequency to pressure

Pressure (Pa)	SAW Peak Frequency (Hz)	Sensitivity(Hz/Pa)
0.87	2,447,989,448	2,807,800,681
2.10	2,448,040,611	41,662.52
7.53	2,448,128,983	16,277.52
20.54	2,448,184,797	4290.37
70.44	2,448,268,518	1677.79
198.86	2,448,338,286	543.25
698.50	2,448,412,705	148.95
2058.71	2,448,487,124	54.71
6849.31	2,448,566,194	16.51
20,070.38	2,448,649,915	6.33
69,677.32	2,448,752,241	2.06
99,722.23	2,448,780,148	0.93

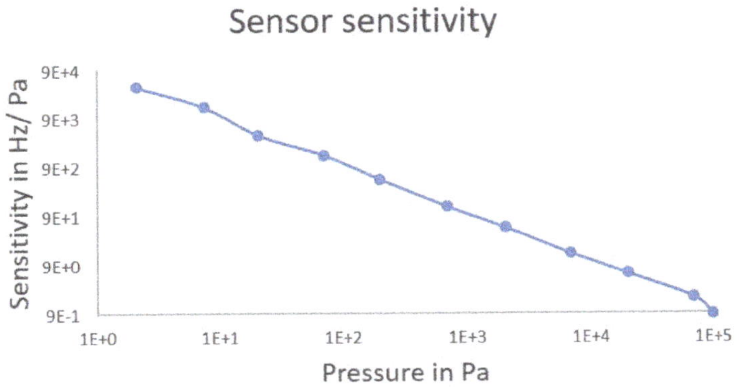

Figure 6. Sensitivity of the SAW chip resonance frequency vs. pressure in Hz/Pa. A logarithmic scale for the pressure axis was used for the sake of clarity.

The values of the gain at SAW peak frequency are similar and do not really depend on the pressure level. The mean value of the gain at the SAW peak frequency is −3.85 dB and the standard deviation is 0.18. Figure 7 shows the values of the gain of the SAW peak versus pressure.

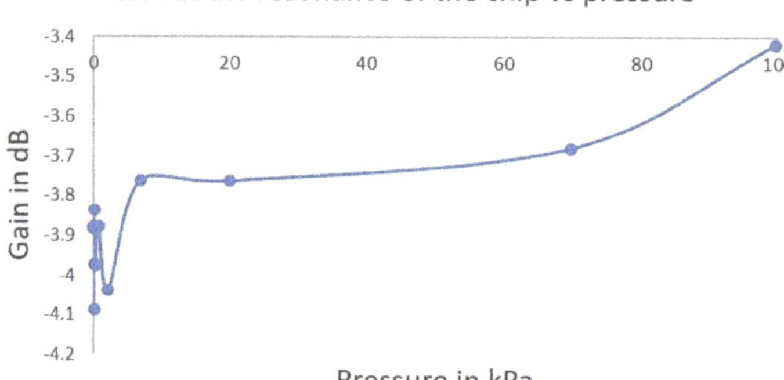

Figure 7. S_{11} gain at the SAW chip resonance frequency vs. pressure in dB.

The values of the gain at the SAW peak frequency are shown in the graph between −4.03776 dB at 2058.71 Pa and −3.417872525 dB at 99,722.23157 Pa. No real trend could be observed despite a continuous increase of pressure. The standard deviation of the gain is stable for each value of the pressure around 0.15%.

The value of the phase at the SAW peak frequency shown in Figure 8 also does not show any obvious trend, have values between 15° and 25° with a mean value of 23.53° and a standard deviation of 3.05°. The values of the phase at peak frequency are situated between 15.87451 at 99,722.23 Pa and 26.36852 at 69,677.32 Pa. The standard deviation of the pressure reported in Table 2 increases with the pressure level, which consolidates the statement of the manufacturer that the uncertainty of the sensor represents 15% of the value of the measurement. The standard deviation of the pressure is always lower than 2% which is much beneath the uncertainty claimed by the manufacturer of the measurement value in any case. The sensitivity decreases when the temperature increases.

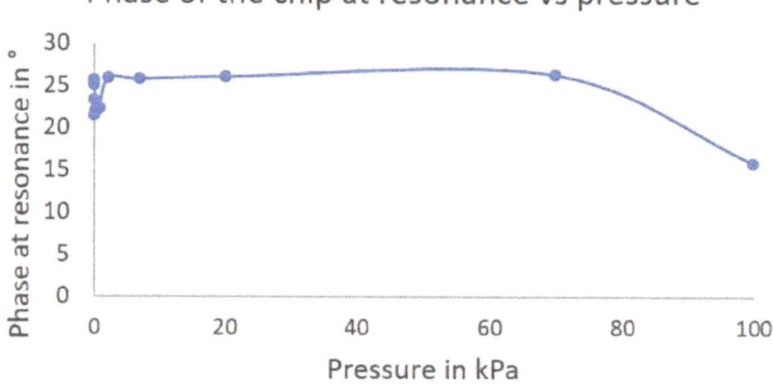

Figure 8. S_{11} phase at the SAW chip resonance frequency vs. pressure in kPa.

4. Discussion

4.1. Sensitivity

The sensitive device is made of lithium niobate Y + 41° − X and has dimensions of 3 mm × 3 mm × 1 mm, which corresponds to the SMD 3 × 3 mm packaging. The SAW wave generated has a temperature coefficient of frequency (TCF) of 19 ppm/K. The equilibrium temperature is 65 °C for a pressure of 0.9 Pa. The heating capacity is 49 mW. Figure 9 shows the equilibrium chip temperature corresponding to each pressure value.

Three regimes can be distinguished in the output curve:

- The Knudsen regime from 0.9 Pa up to 200 Pa approximately, where the highest sensitivity is reached.
- The transition regime from 200 Pa up to 7000 Pa approximately where the sensitivity takes intermediate values.
- The convection regime approaching atmospheric pressure where the sensitivity is even lower than in transition regime in the range of 1 Hz/Pa.

The sensitivity rapidly decreases when the pressure increases. This was expected. It is due to the exit of the Knudsen regime as pressure increases and the corresponding evolution of the gas thermal conductivity towards a constant asymptotic value. The mean free path of the gas molecules at ambient temperature decreases very quickly between 0.01 and 100 Pa from 0.1144 m to 0.0011444 m which corresponds to a Knudsen number from 0.715 to 7.15. Considering the dimensions of the chamber whose diameter is 0.160 m, the pressure range tested corresponds to the transition from the Knudsen regime to a regime with a constant thermal conductivity hence the observed responses. The experimental curve of the sensitivity versus pressure is shown in Figure 6. It was obtained by deriving the values of resonance frequency with respect to pressure. The equilibrium temperature of the chip for each pressure is shown in Figure 9. The equilibrium temperature was computed from the value of the resonance frequency using the temperature coefficient of frequency provided in the technical sheet of the chip which is 18.857 ppm/K.

The response of the sensor to the pressure is non linear. First of all the Pirani sensing curve has a characteristic S-shape [8] due to the nonlinear relationship between the gas thermal conductivity and pressure. However, eliminating thermal losses, eliminating the offset of the signal and applying a proper algorithm to the output signal may significantly increase the sensor linearity.

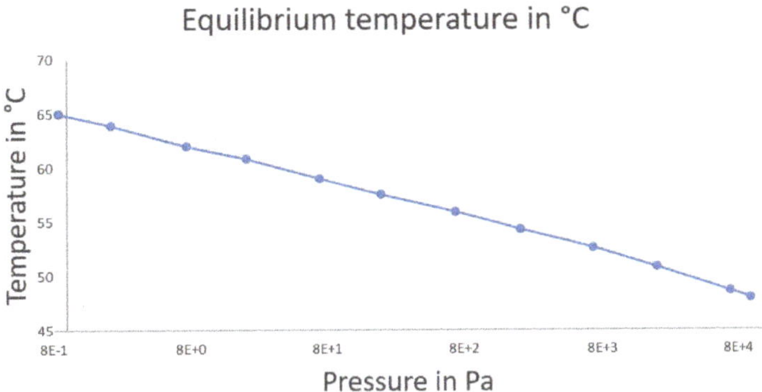

Figure 9. Temperature of the chip versus pressure.

The experimental points obtained are compatible with the theoretical expectations in their trend. This means that the SAW peak frequency increases when the pressure increases with a higher sensitivity in the fine vacuum region between 0.1 and 1000 Pa than between 1000 and 100,000 Pa in the rough vacuum area. This was to be expected from the Knudsen regime analysis, since the thermal conductivity tends towards a constant value approaching atmospheric pressure.

Another less impactfull phenomenon can explain the decrease of sensitivity when the pressure increases. The sensitive device cools down when the pressure increases. The sensitivity of the device to the pressure decreases when its temperature approaches the ambient temperature. This phenomenon is nonetheless less influential than the exit of the Knudsen regime.

4.2. Uncertainty

Every measurement is subject to measurement uncertainty, which is the statistical deviation of the measured values from their true value. There are two types of measurement uncertainty: random uncertainty and systematic uncertainty. Random uncertainties vary with time and are thus unpredictable. While they can be described by statistical analysis, they cannot be removed by calibration. Common random uncertainties include those related to instrument noise and the repeatability of switches, cables, and connectors. On the contrary, systematic uncertainties occur in a reproducible way. They are caused by imperfections in the VNA and can be taken into account by means of calibration.

The severity of the random uncertainty can be mitigated by adopting careful and meticulous measurement practices, such as allowing the instrument to achieve thermal equilibrium, using high-quality cables and connectors, selecting a small frequency bandwidth, and using averaging. This was attempted by working in a temperature controlled environment, using thick cables less vulnerable to bending, working on a restricted frequency range corresponding to the location of the SAW signal and using averaging algorithms. A VNA is also sensitive to its surrounding temperature, which can cause a drift. This was addressed by working in a temperature controlled lab with 1 °C drift.

An uncertainty analysis and a statistical analysis were performed. The error bars were computed based on the standard deviation (shown in Table 2). The sample mean value and standard deviation were computed.

This uncertainty remains acceptable given the uncertainties on each system parameter (pressure, TCF, and heating power). The agreement between the theoretical curve and the experimental response is within a range of +/−20%. This allows one to validate the dynamic model of the SAW-Pirani developed. Variations of 1 Pa can be clearly detected. A zoom on the noisiest SAW peak makes it possible to estimate the limit step of measurement at about 1.5 kHz.

Table 2. Sensitivity of the prototype.

Pressure (Pa)	Pressure Standard Deviation	Sigma/P(%)
0.87	0.0020	0.22
2.10	0.0136	0.65
7.53	0.1028	1.37
20.54	0.4161	2.03
70.44	0.1681	0.24
198.86	1.0710	0.54
698.50	5.0648	0.73
2058.71	16.0690	0.78
6849.31	16.9220	0.25
20,070.38	15.0446	0.075
69,677.32	115.3257	0.17
99,722.23	126.8275	0.13

The experimental points are compatible with the theoretical expectations. The similarity between theory and experimental results validates the theoretical model to compute the sensitivity of a SAW-Pirani sensor. The precision of the reference sensor is not constant and depends on the pressure range. The sensitivity decreases continuously through the pressure range. The uncertainty of the gauge is 15,000 Pa close to atmospheric pressure and 1.5×10^{-5} Pa in a strong vacuum. This diminution also accounts for the discrepancy observed between the experimental results and the theoretical curve computed for a gas in the Knudsen regime. The theoretical curve corresponds to the maximum sensitivity. It was therefore expected that experimental results would be of a lower quality. There are also other sources of uncertainty among which the equilibrium temperature that depends on the pressure, the ambient temperature and the input voltage fluctuations.

No frequency counter was used since there was more interest in the shift than in the actual value itself. The basis frequency depends on the thermal resistance and the size of the device.

However, no extended uncertainty analysis was carried on. The analysed system is a prototype, many practical aspects were validated and a qualitative result was expected: the trend of the resonance frequency versus pressure. Extensive uncertainty analysis should be performed only after optimisation of the prototype and optimised integration inside vacuum.

4.3. Signal Processing

In the Pirani sensing the thermal losses are conduction losses and then gaseous thermal transfer losses. Since the gaseous heat transfer is faster, the response time decreases. When AC current is applied, the response time generates a phase between the input signal and the temperature rise. It is easy to observe by performing a frequency sweep to the nanowire. When the pressure in the chamber is modified, the time constant is modified and the temperature response is shifted. The quasi-static approach is no longer valid. The nanowire and the gas gap are considered as thermal resistances and capacitances in parallel and subject to heat flux. When the nanowire is transferred from vacuum to gas medium, the thermal resistance decreases which lowers the response time.

A method to reduce noise is to reduce the VNA's intermediate frequency (IF) bandwidth. The IF bandwidth is a digitally implemented variable filter used to reduce noise. Narrower IF BW's result in additional data samples at each frequency point sample. The effect is similar to averaging. However, IF bandwidth reduction results in an increase in point-to-point averaging. Normal averaging, as noted earlier, is a sweep-by-sweep or trace-by-trace average. The minimum measurable frequency step is 1 kHz, this is the sensitivity of the network analyser. Measurement performance depends on the calibration standard chosen, and on the repeatability of the measurement system itself.

A visual criterion was used to identify the SAW peak, an algorithm is needed with clear conditions to identify the SAW peak. These points can be compared to the curve obtained with the simulation of 100 ppm/K calculated with a surface emissivity of 0.75. The experimental values follow the expected trend but are quite far from the orders of magnitude expected.

4.4. Comparison to Other Works

In [10], to determine the sensitivity of the device close to the Knudsen regime, the ultimate pressure of the pump was maintained during the whole experiment. The power injected in the heating resistor was increased by threshold. For each threshold, after the thermal equilibrium was reached and the temperature computed, the pressure was variated by a few tenths of Pa. The relative frequency shift generated by this threshold was measured and the ultimate pressure reinstalled. The sensitivities of the device for each threshold were extracted.

In [12], several low pressure gauges are compared. An uncertainty analysis is presented. The most important source of noise identified was the short term random noise. The short term random noise creates instabilities in zero-pressure readings. Another source of uncertainty is, for heated gauges, thermal transpiration at absolute pressures below 100 Pa. At higher pressures performance is limited by long-term shifts in calibration with time (months to years). Random noise limits the smallest pressure change that can be resolved by a transducer. A measure of the noise-limited pressure resolution is given by twice the standard deviation of repeated readings at a stable pressure. The resolution of different transducers tends to scale linearly with their full scale range. The capacitive diaphragm gauges (CDGs) have a resolution of about 1 part in 106 of full scale (FS) range, quartz based transducers about 1 part in 106 to 3 parts in 106 of FS and MEMS-type transducers at about 4 parts in 106 to 10 parts in 106 of FS. As of their availability with lower FS ranges, CDGs have the best absolute pressure resolution among the transducers. The zero instabilities in transducers manifest themselves primarily in two ways. They appear either as zero shifts that correlate directly with changes in room temperature, or as zero drifts that vary randomly in both sign and magnitude and are probably due to drifts in electronics and/or mechanical structure of the gauge.

In [13], a stable resonant pressure sensor fabricated using 3-D micromachining process is described. Two resonators are located on the surface of a diaphragm and applied pressure is measured from the difference of two resonant frequencies. The sensors is located in a 6.8 × 6.8-mm wide, 0.5-mm thick silicon chip. The resonator has a high Q-value of 50,000. The principle of a resonant pressure sensor is easily understood by an analogy of the phenomenon in which a natural frequency of a string changes due to tensile force. EPROM in the pressure receiver stores the sensor parameters for signal linearization and wide range temperature compensation.

A PhD thesis by Ruellan [14] presents many interesting conclusions. Ruellan proposes to use the transient regime to reduce response time of thermal conductivity pressure gauges. Using AC signal increases the sensitivity by reducing the noise. Reducing the noise also comes with a reduction of drifts. This can be seen by applying pressure thresholds to the sensing device. The omega method is used to determine thermal conductivity. The small response time can be measured by applying a step current to a nanowire and observing the voltage output. A simple nanowire has a resistance in the range of kΩ. The capacitance of a coaxial cable is in the range of 100 pF. For a 1 m cable of 20 Ω, the characterstic response time is 2 μs. This time is close to the thermal response time of the device. The electric filtering may disturb the measurement. A solution to this can be the use of parallel nanowires to reduce the overall resistance. Coaxial cables tend to be replaced by twisted pairs that have a lower lineal capacitance. Short pairs are preferred to reduce the capacitance to the maximum. The cabling used for signal acquisition impacts the response time in the μs range. The response time depends on pressure.

4.5. Outlook

Considering all the limitations and adjustments currently required, it can be asked whether the designed sensor and prototype represents an advantageous solution for the future.

The commercially available viewports do not allow the transmission of either waves. Besides, to prevent tensions occurring during heating, cooling or installing, the viewports contain an intermediate metallization layer of kovar (an iron-nickel-cobalt alloy) that prevents electromagnetic waves propagation. PMMA is not the best material to use in vacuum due to its poor mechanical stability. When vacuum was applied it was systematically bending which presented a risk of failure for such a small thickness. Viewports of ceramics without metallization or quartz should be considered for further tests. In [10] a fused quartz window is used to perform transvacuum measurements. Other distances due to the interrogation antenna and PCB were introduced. The reference [15] highlights the wireless charging procedure of a metallic device. It addresses the issue of electromagnetic waves transmission in metals. In vacuum most equipment is in stainless steel or aluminum, using the surrounding metals as an intermediate coil for induction coupling could be a solution. If this method can also be transposed to antenna operation, there will be no need for windows anymore.

The sensitivity and repeatability of the measurements were assessed. The presettings of the VNA software influences the quality of the measurement signal without modifying the hardware structure. A VNA of better signal quality and denser frequency sweep should be used to obtain a better precision. Since the prototype tested is not a final product, there is no use of an extended calibration according to international standards yet. For this purpose, a pressure calibrator (AMETEK JOFRA AP010C, AMETEK, Inc., Berwyn, PA, USA) [16] could be used.

The final setup in vacuum aimed to operate the prototype wirelessly and see its response. The position of each of the components strongly influences the measurement signal. All the components impact the measurements. In [17–19], an ultrasonic interferometer is presented. Factors for accurate low pressure manometry include the control of thermal and mechanical instabilities.

The sensor itself shall have higher performances thanks to successful manufacturing processes. In order to enhance the performances of the prototype, many improvements in the manufacturing should be made. The SAW device should directly be suspended by wires to the substrate to eliminate the solid conduction heat losses. The heating resistance should be etched around the IDT directly in contact with the gas during the lithography process prior to the dicing.

The PCB containing the heating and interrogation unit should also be adapted to support the suspended chip by containing a hollow at its center aligned to the position of the microchannel. The SAW chip of the sensor is made of lithium niobate Y + 41°-X. However, using the YZ cut of the same lithium niobate crystal increases the temperature sensitivity up to 100 ppm/K which increases the temperature sensitivity by approximately five times. The results presented so far can be compared to the theoretical sensitivity curve mentioned above. For this purpose the experimental sensitivities should be multiplied by a correction factor to take the TCF difference into account. This factor equals to 100/19 = 5.3 approximately. In order to improve the SAW Pirani performances, a more robust and easier to manufacture prototype should be produced. The IDTs can be in Ti/Au or in Ta/Pt. The Ti/Au electrodes can be used at low temperature. A diffusion phenomenon followed by the oxydation of titanium in gold at higher temperature forbids their use beyond 300 °C. The Ta/Pt structure is not subject to this issue and can be operated at very high temperatures. The use of a Pt heating resistor enables the direct measurement of the substrate temperature with conventional Pt 100 gauge for instance.

A dynamic model needs to be validated now to forecast the response time of the sensor with respect to its geometry and its thermal resistance. It might then be possible to detect 0.0001 Pa. Finally, the sensitivity can be increased and the limit step further reduced by metallizing the surface of the sensing element. The experimental results also confirm the theoretically planned measurement range. Figure 5 shows the response of the SAW-Pirani over a wide range of pressure values, without forced convection device. Further experiments would have to show whether it is possible to remove the zero sensitivity area (regime 2) by means of a forced convection device.

5. Conclusions

A prototype of a compact wireless SAW Pirani vacuum sensor was manufactured and tested. Measurements inside vacuum betwwen 0.9 Pa and atmospheric pressure allowed to test the complete prototype inside vacuum with its components—the sensing chip, the heating unit, the interrogation unit, the PCB and the packaging. Completely wireless measurements were performed via induction coupling and antenna reflection in order to observe a pressure-dependent behaviour of the chip.

There is still room for improvement in almost every aspect, mainly the vacuum interface, the electrical contacting of the chip and the antenna coupling. However, the technical solution imagined to wirelessly measure vacuum pressure with a compact sensor proved to be feasible.

Author Contributions: Conceptualization, S.T. and J.J.B.; investigation, S.S., H.D. and M.J.; data curation, A.V. and S.T.; writing—original draft preparation, S.T.; writing—review and editing, J.J.B. and J.G.K.; supervision, J.J.B.; project administration, J.J.B.; funding acquisition, J.J.B. All authors have read and agreed to the published version of the manuscript.

Funding: This research received no external funding.

Acknowledgments: The authors would like to acknowledge the financial support provided by the EU network program H2020 MIGRATE under grant number 643095. The authors also acknowledge support by the KIT-Publication Fund of the Karlsruhe Institute of Technology.

Conflicts of Interest: The authors declare that they have no conflict of interest.

References

1. Fu, C.; Ke, Y.; Li, M.; Luo, J.; Li, H.; Liang, G.; Fan, P. Design and Implementation of 2.45 GHz Passive SAW Temperature Sensors with BPSK Coded RFID Configuration. *Sensors* **2017**, *8*, 1849. [CrossRef]
2. Pohl, A. A review of wireless SAW sensors. *IEEE Trans. Ultrason. Ferroelectr. Freq. Control* **2000**, *47*, 317–332. [CrossRef]

3. Nicolay, P.; Lenzhofer, M. A Wireless and Passive Low-Pressure Sensor. *Sensors* **2014**, *14*, 3065–3076. [CrossRef] [PubMed]
4. Singh, K.J. Enhanced sensitivity of SAW-based pirani vacuum pressure sensor. *IEEE Sens. J.* **2011**, *11*, 1458–1464. [CrossRef]
5. Nicolay, P., Elmazria, O.; Perois, X. A miniaturized SAW-PIRANI Sensor. In Proceedings of the IEEE International Ultrasonics Symposium transaction, Dresden, Germany, 7–10 October 2012.
6. Mueller, A.; Konstantinidis, G.; Giangu, I.; Adam, C.G.; Stefanescu, A.; Stavrinidis, A.; Stavrinidis, G.; Kostopoulos, A.; Boldeiu, A.; Dinescu, A. GaN Membrane Supported SAW Pressure Sensors with Embedded Temperature Sensing Capability. *IEEE Sens. J.* **2017**, *17*, 7383–7393. [CrossRef]
7. Borrero, G.A.; Bravo, J.P.; Mora, S.F.; Velásquez, S.; Segura-Quijano, F.E. Design and fabrication of SAW pressure, temperature and impedance sensors using novel multiphysics simulation models. *Sens. Actuators A Phys.* **2013**, *203*, 204–214. [CrossRef]
8. Toto, S.; Nicolay, P.; Morini, G.L.; Rapp, M.; Korvink, J.K.; Brandner, J.J. Design and simulation of a wireless sawpirani sensor with extended range and sensitivity. *Sensors* **2019**, *19*, 2421–2441. [CrossRef]
9. Toto, S.; Nicolay, P.; Morini, G.L.; Voigt, A.; Korvink, J.G.; Brandner, J.J. Towards a compact wireless surface acoustic wave Pirani microsensor with extended range and sensitivity. *Heat Transf. Eng.* **2020**, *42*. [CrossRef]
10. Nicolay, P. Les Capteurs à ondes éLastiques de Surface: Applications Pour la Mesure des Basses Pressions et des Hautes TempéRatures. Ph.D. Thesis, Université Henri Poincaré, Nancy, France, 2007.
11. Kandlikar, S.; Garimella, S.; Li, D.; Colin, S.; King, M. *Heat Transfer and Fluid Flow in Minichannels and Microchannels*, 2nd ed.; Elsevier: Oxford, UK, 2014.
12. Mueller, A.P. Measurement performance of high-accuracy low-pressure transducers. *Metrologia* **1999**, *36*, 617–621. [CrossRef]
13. Harada, K.; Ikeda, K.; Kuwayama, H.; Murayama, H. Various applications of resonant pressure sensor chip based on 3-D micromachining. *Sens. Actuators A Phys.* **1999**, *73*, 261–266. [CrossRef]
14. Ruellan, J. Design, fabrication and characterization of a silicon nanowire based thermal conductivity detector. In *Archives Ouvertes*; Université Grenoble Alpes: Grenoble, France, 2015.
15. Jeong, N.S.; Carobolante, F. Wireless Charging of a Metal-Body Device. *IEEE Trans. Microw. Theory Tech.* **2017**, *4*, 1077–1086. [CrossRef]
16. Rodríguez-Madrid, J.G.; Iriarte, G.F.; Williams, O.A.; Calle, F. High precision pressure sensors based on SAW devices in the GHz range. *Sens. Actuators A Phys.* **2013**, *189*, 364–369. [CrossRef]
17. Heydemann, P. L.M.; Tilford, C.R.; Hyland, R.W. Ultrasonic manometers for low and medium vacua under development at the National Bureau of Standards. *J. Vac. Sci. Technol.* **1977**, *14*, 597–605. [CrossRef]
18. Greve, D.W.; Chin, T.L.; Zheng, P.; Ohodnicki, P.; Baltrus, J.; Oppenheim, I.J. Surface acoustic wave devices for harsh environment wireless sensing. *Sensors* **2013**, *13*, 6910–6935. [CrossRef] [PubMed]
19. Mercier, D.; Bordel, G.; Brunet-Manquat, P.; Verrun, S.; Elmazria, O.; Sarry, F.; Belgacem, B.; Bounoua, J. Characterization of a SAW-Pirani vacuum sensor for two different operating modes. *Sens. Actuators A Phys.* **2012**, *188*, 41–47. [CrossRef]

Publisher's Note: MDPI stays neutral with regard to jurisdictional claims in published maps and institutional affiliations.

Article

VectorDisk: A Microfluidic Platform Integrating Diagnostic Markers for Evidence-Based Mosquito Control

Sebastian Hin [1,†], Desirée Baumgartner [1,2,†], Mara Specht [1,†], Jan Lüddecke [1], Ehsan Mahmodi Arjmand [1], Benita Johannsen [1], Larissa Schiedel [1], Markus Rombach [1], Nils Paust [1,2], Felix von Stetten [1,2], Roland Zengerle [1,2], Nadja Wipf [3,4], Pie Müller [3,4], Konstantinos Mavridis [5], John Vontas [5,6] and Konstantinos Mitsakakis [1,2,*]

1 Hahn-Schickard, Georges-Koehler-Allee 103, 79110 Freiburg, Germany; sebastian.hin@iuvas.de (S.H.); Desiree.Baumgartner@imtek.uni-freiburg.de (D.B.); Mara.Specht@Hahn-Schickard.de (M.S.); Jan.Lueddecke@Hahn-Schickard.de (J.L.); Ehsan.Arjmand@Hahn-Schickard.de (E.M.A.); Benita.Johannsen@Hahn-Schickard.de (B.J.); Larissa.Schiedel@imtek.uni-freiburg.de (L.S.); Markus.Rombach@Hahn-Schickard.de (M.R.); Nils.Paust@Hahn-Schickard.de (N.P.); Felix.von.Stetten@Hahn-Schickard.de (F.v.S.); Roland.Zengerle@Hahn-Schickard.de (R.Z.)

2 Laboratory for MEMS Applications, IMTEK—Department of Microsystems Engineering, University of Freiburg, Georges-Koehler-Allee 103, 79110 Freiburg, Germany

3 Swiss Tropical and Public Health Institute, Socinstrasse 57, 4002 Basel, Switzerland; nadja.wipf@swisstph.ch (N.W.); pie.mueller@swisstph.ch (P.M.)

4 Department of Epidemiology and Public Health, University of Basel, Petersplatz 1, 4001 Basel, Switzerland

5 Foundation for Research and Technology-Hellas, Institute of Molecular Biology and Biotechnology, 70013 Heraklion, Greece; mavridiskos@gmail.com (K.M.); vontas@imbb.forth.gr (J.V.)

6 Pesticide Science Laboratory, Department of Crop Science, Agricultural University of Athens, 11855 Athens, Greece

* Correspondence: konstantinos.mitsakakis@imtek.uni-freiburg.de or Konstantinos.Mitsakakis@Hahn-Schickard.de; Tel.: +49-761-203-73252

† These authors have equally contributed.

Received: 24 November 2020; Accepted: 15 December 2020; Published: 18 December 2020

Abstract: Effective mosquito monitoring relies on the accurate identification and characterization of the target population. Since this process requires specialist knowledge and equipment that is not widely available, automated field-deployable systems are highly desirable. We present a centrifugal microfluidic cartridge, the VectorDisk, which integrates TaqMan PCR assays in two feasibility studies, aiming to assess multiplexing capability, specificity, and reproducibility in detecting disk-integrated vector-related assays. In the first study, pools of 10 mosquitoes were used as samples. We tested 18 disks with 27 DNA and RNA assays each, using a combination of multiple microfluidic chambers and detection wavelengths (geometric and color multiplexing) to identify mosquito and malaria parasite species as well as insecticide resistance mechanisms. In the second study, purified nucleic acids served as samples to test arboviral and malaria infective mosquito assays. Nine disks were tested with 14 assays each. No false positive results were detected on any of the disks. The coefficient of variation in reproducibility tests was <10%. The modular nature of the platform, the easy adaptation of the primer/probe panels, the cold chain independence, the rapid (2–3 h) analysis, and the assay multiplexing capacity are key features, rendering the VectorDisk a potential candidate for automated vector analysis.

Keywords: microfluidics; LabDisk; vector-borne diseases; malaria; arboviruses; insecticide resistances; mosquito monitoring

1. Introduction

Vector-borne diseases make up more than 17% of all infectious diseases globally, and are responsible for more than 700,000 deaths annually, according to the World Health Organization (WHO) [1]. Vectors are arthropod species (e.g., mosquitoes, ticks, and sandflies) that have the capability of acquiring and transmitting at least one type of vector-borne diseases, for example malaria (parasitic disease), dengue, chikungunya, West Nile, Zika, and yellow fever (viral diseases). Malaria is the most prevalent of all, with an estimated 220 million global cases resulting in more than 400,000 deaths annually, while dengue is the most prevalent viral infection, with an estimated 96 million global cases resulting in more than 40,000 deaths annually [1]. Such diseases often appear as epidemic outbreaks, thereby disrupting the countries' healthcare and socioeconomic systems [2].

Although mosquito-borne diseases have typically been a problem in tropical countries, nowadays they also represent a threat in regions with more temperate climates [3,4]. Indicative examples are: (i) the spread of the invasive Asian tiger mosquito (*Aedes albopictus*), which has been associated with disease transmission in north-western USA [5] and in Europe (in the latter, specifically, it has been associated with several local dengue [6], chikungunya [7,8], and Zika virus [9] outbreaks); and (ii) the continuous circulation of West Nile virus, vectored by the common mosquito *Culex pipiens* (almost unnoticed until recently), which has caused epidemics with fatalities in several European countries [10–12].

Vector control relying primarily on the use of insecticides has successfully reduced disease cases globally. Since 2000, malaria has been reduced by almost 50% due to insecticide-treated nets (ITN) and indoor residual spraying (IRS) in combination with case management with artemisinin-based drugs [13]. However, the successes are being challenged by (i) increasing insecticide resistance [10,14,15], and (ii) current tools being ineffective by targeting the wrong vectors. For example, ITN and IRS do not target outdoor biting malaria vectors. Therefore, reliable information on the target vector is crucial, and suitable, integrated, and automated vector analysis tools are desirable to make evidence-based vector control more effective [16].

Three levels of information are crucial for vector surveillance and control: (i) mosquito species identity is essential for differentiating vector from non-vector species (e.g., out of 476 *Anopheles* species, only 70 are capable of transmitting malaria) [17]; (ii) pathogen detection in the case of malaria vectors, the determination of the *Plasmodium* species (*P. falciparum*, *P. vivax*, *P. ovale*, or *P. malariae*), whether the parasites are present in their infective stage [18,19], and—in the case of vectors of viral diseases—the determination of arbovirus species (e.g., West Nile virus, dengue virus); and (iii) the potential involvement of an insecticide resistance mechanism, including target site mutations (DNA) and expression levels of metabolic detoxification genes (RNA) [20,21]. Despite various methods that have been developed in the past to address one or more of these characteristics [22], there is no universal tool to collect all necessary information at once; the individual protocols can be very tedious and can involve manual steps such as the dissection of salivary glands [23], or require the analytical skills to perform microscopic identification through morphological examination.

A recent collaborative effort by the authors has resulted in the development and optimization of multiplex TaqMan probe-based reverse transcription polymerase chain reaction (RT-PCR) assays to acquire information on the aforementioned three levels [24]. The current work is a follow-up intended to integrate these TaqMan assays into a microfluidic platform, enabling their automated realization. The integration is achieved on a centrifugal microfluidic cartridge, the VectorDisk, which is an application-specific version of the LabDisk platform. The LabDisk has been shown to operate with pathogen detection from human samples [25–27], but this was typically in a single- or two-target configuration (one or two pathogens simultaneously present in a sample in the form of co-infections). The current work aims to demonstrate the capabilities of the platform in: (i) simultaneously acquiring results from multiple DNA and RNA assays using geometric and color multiplexing (namely multiple microfluidic reaction chambers and detection wavelengths, respectively); (ii) using diluted crude lysates of pools of 10 whole mosquitoes as starting material without the need for exhaustive nucleic acid purification; and (iii) easy adaptation from *Anopheles*- to *Aedes*-species-specific assay detection. We assessed the feasibility and reproducibility of the assay integration onto the disk in two small-scale studies: the Vector Study, starting from laboratory-reared mosquitoes, to test the malaria assays; and the Nucleic Acid (NA) Study, starting from purified nucleic acids, to test the arbovirus assays [28]. This proof-of-principle work indicates the potential of the platform to be used for fully automated analysis in the framework of integrated vector control.

2. Materials and Methods

2.1. Mosquito Assays

In order to demonstrate the capability of the VectorDisk platform to simultaneously collect different molecular information for species identification and insecticide resistance, the assays used in this work were designed to detect: (i) DNA-based mosquito species of the *Anopheles gambiae* species complex; (ii) if present, the DNA of the malaria parasite, thereby differentiating between *P. falciparum* (the most lethal for humans) and *P. ovale/vivax/malariae* [18], and the infective stage through the RNA detection of sporozoite-specific *P. falciparum* transcripts [29]; (iii) knockdown resistance (kdr) point mutations conferring resistance in the target site of the para voltage-gated sodium channel responsible for the insensitivity to dichlorodiphenyltrichloroethane (DDT) and pyrethroids [30–32], as well as target site resistance to organophosphates and carbamates due to insensitive acetylcholinesterase (iAChe) [33]; (iv) overexpression of detoxification loci at the mRNA level [34], which can pinpoint associations with insecticide resistance, such as the cytochrome P450 monooxygenases CYP6P3, CYP6M2, CYP6P4, and CYP6Z1, and the glutathione-S transferase GSTE2 in *An. gambiae*; (v) the viral RNA, which may be present in arbovirus vectors (i.e., West Nile virus [35] and Zika virus [36]).

2.2. Vector Study: Manual Preparation of Mosquito Samples (Lysis and Homogenization)

The insectary at the Swiss Tropical and Public Health Institute provided specimens of the Kisumu colony that originates from an MRA-762 egg batch provided by BEI Resources. For this Vector Study, 50 mosquitoes (Kisumu, 25 male (m) + 25 female (f)) were used, and the following preparatory steps were performed prior to the automated VectorDisk protocol. The mosquitoes were split into five pools of ten mosquitos each (5 males + 5 females). It was shown by Mavridis et al. [31] that the use of pools of ten mosquitoes saves time, reduces costs, and allows for more efficient resistance allele screening than in the case of individual mosquito testing. The mosquitoes were preserved in RNAlater (Thermo Fisher Scientific, Waltham, MA, USA), and thawed on a lint-free towel. They were then washed with ultra-pure H_2O to remove the RNAlater. Subsequently, they were ground manually in a 150 μL lysis buffer (RLT Buffer, QIAGEN, Hilden, Germany) with a pestle (on ice). The homogenate was incubated for 10 min at room temperature. After centrifugation for 2 min at 16,000 rcf, the supernatant was collected and stored at −20 °C in aliquots. Prior to any next experimental step, the lysate was diluted 1:200 with ultra-pure H_2O. In order to acquire results that would allow some reproducibility assessment, one lysate from each pool was diluted and aliquoted for testing three to four times. Thus, we could perform inter-pool as well as intra-pool reproducibility assessments. A summary is shown in Figure 1. A total of 18 disks were used in the Vector Study.

2.3. Nucleic Acid Study

The study includes the following purified genomic RNA that was obtained through BEI Resources, NIAID, and NIH: genomic RNA from West Nile virus (WNV), Bird 114, NR-9573, genomic RNA from Zika virus (ZIKV), FLR-Zika virus, NR-50241. The RNA concentration was <1.0 ng/μL. For testing the *Plasmodium* and the infected stage assays (sample coded as 'PF-INF'), we included in the NA Study a mixture of DNA and RNA extracted from laboratory-reared *P. falciparum* infected mosquitoes as previously described [29] and their concentration was 120 ng/μL. In order to acquire results that would allow some reproducibility assessment, each sample was tested three times in the VectorDisk.

As each VectorDisk included each primer/probe set twice, we acquired six data points for each assay. A summary is shown in Figure 2. A total of nine disks were used in the NA Study.

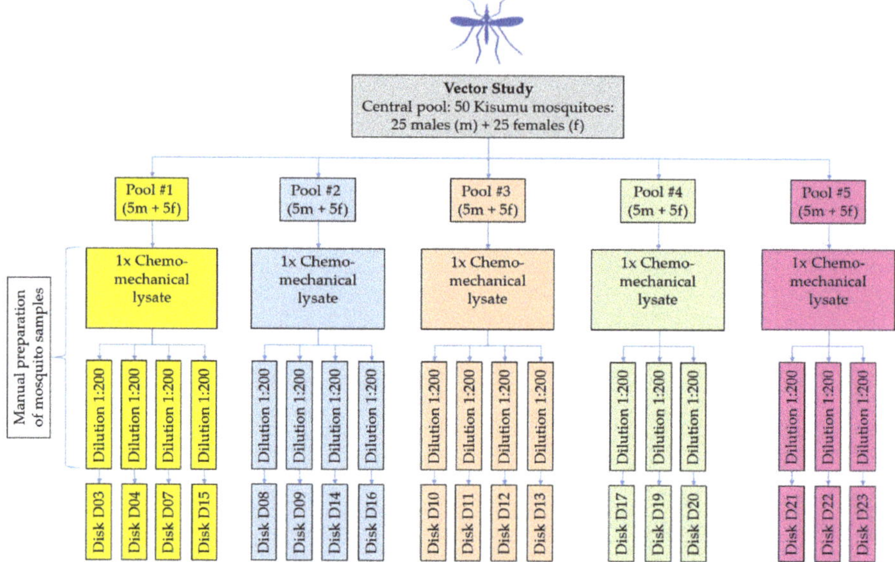

Figure 1. Schematic workflow for manual preparation of mosquito samples prior to VectorDisk tests during the Vector Study.

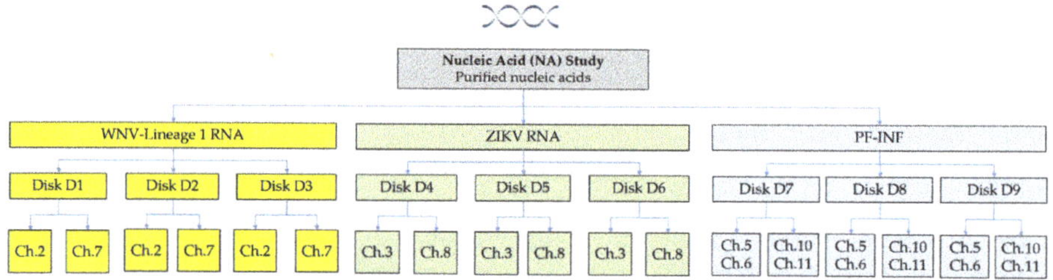

Figure 2. Schematic workflow of testing the three nucleic acid (NA) samples in the VectorDisk during the NA Study. 'PF-INF' represents the mixture of *P. falciparum* DNA + *P. falciparum* infective stage RNA.

2.4. VectorDisk Fabrication

The microfluidic design was created using SolidWorks2017 (Dassault Systèmes SolidWorks Corp., Waltham, MA, USA). Production of the VectorDisk cartridge was done by Hahn-Schickard Lab-on-a-Chip Foundry [37] using rapid tooling and a scalable and low cost micro-thermoforming process for replication of prototypes [38,39] made of Cyclic Olefin Copolymer foil (COC 6013/8007, 200 µm thick, TOPAS Advanced Polymers GmbH, Frankfurt, Germany). For this, an in-house milled (EVO, Kern Microtechnik GmbH, Eschenlohe, Germany) metal master was used on a batch production thermoforming machine (Rohrer AG, Möhlin, Switzerland). No further surface treatment was performed after thermoforming. The primers/probes, together with trehalose as stabilizer (final concentration 50 mM), were pipetted in the reaction chambers, and the cartridges were then left to dry in an oven at 50 °C for 1 h. The combinations of the TaqMan probe assays are shown in Tables 1 and 2. Details on the primer/probe design and validation of the assays, including sensitivity results, are available in previous studies for the malaria-specific [29,31] and arboviral-specific [35,36,40] assays. The primer/probe sequences are given in Supplementary Tables S1 and S2. The structured foil was sealed with a 200 μm thick COC foil (COC 6013/8007, TOPAS Advanced Polymers GmbH, Germany) using an adaptive thermo-sealing process [41] on a modified hot embossing machine (Wickert GmbH, Landau in der Pfalz, Germany). The sealing foil had an opening at the radius that corresponded to the mixing chamber location, where the lyopellet was inserted manually after the thermal sealing. The lyopellets were developed and supplied specifically for these triplex assays (Fast Track Diagnostics, Luxembourg), and their functionality was tested and validated prior to their use in this work [34]. At the end, a piece of pressure-sensitive adhesive foil 9795R (3M, Maplewood, MN, USA) was used to seal the lyopellet mixing chamber. The cartridges were stored with a desiccant in a sealed aluminium pouch to protect reagents from humidity and light until use.

Table 1. VectorDisk assay panel for the Vector Study (triplex or duplex, depending on the assay of the VectorDisk). Targets are detected in 11 different reaction chambers and in three different detection channels (FAM-green, HEX-yellow, and ATTO 647N-red), detecting the fluorescence signal from the corresponding fluorophores in brackets.

Reaction Chamber	Assay Panel	Nucleic Acid Type	Assay Target (FAM)	Assay Target (HEX)	Assay Target (ATTO 647N)
2	Species ID	DNA	Aq [1]	Ag [1]	Aa [1]
3	Molecular Forms	DNA	S [2]	M [2]	-
4	Kdr	DNA	Rw [3]	S-wt [4]	Re [3]
5	Kdr+	DNA	R	S-wt	-
6	iAChe	DNA	R	S-wt	-
7	*Plasmodium* species	DNA	*P. falciparum*	*P. ovm* [5]	-
8	Infective stage	RNA	-	PLP1	-
9	Detox (A)	RNA	RPS7	CYP6P3	CYP6M2
10	Detox (B)	RNA	RPS7	CYP9K1	CYP6P4
11	Detox (C)	RNA	RPS7	CYP6Z1	GSTE2
12	Detox (D)	RNA	RPS7	CYP6P1	CYP4G16

[1] Aq: *Anopheles quadriannulatus/merus/melas/bwambae*, Ag: *An. gambiae* s.s. (sensu stricto)/*An. coluzzii*, Aa: *An. arabiensis*, [2] S: *An. gambiae* s.s. (former molecular S-form), M: *An. coluzzii* (former molecular M-form), [3] Rw: Resistant allele west, Re: Resistant allele east, [4] S-wt: Susceptible (wild type), [5] *P. ovm*: *P. ovale/vivax/malariae*, i.e., the primer set detects all three. The symbol '-' means that there were no assay targets expected to be detected in the particular reaction chambers and detection channels.

Table 2. VectorDisk assay panel for the NA Study. Targets are detected in 10 different reaction chambers and in three different detection channels (FAM-green, HEX-yellow, and ATTO 647N-red), detecting the fluorescence signal from the corresponding fluorophores in brackets.

Reaction Chamber	Assay Panel	Nucleic Acid Type	Assay Target (FAM)	Assay Target (HEX)	Assay Target (ATTO 647N)
2	West Nile virus	RNA	WNV-Lineage 1	WNV-Lineage 2	-
3	Zika virus	RNA	-	-	ZIKV
4	Dengue virus	RNA	DENV T1-4	-	-
5	*Plasmodium* species ID	DNA	*P. falciparum*	*P. ovm* [1]	-
6	*Plasmodium* infective stage	RNA	-	-	*Pf* infective stage
7	West Nile virus	RNA	WNV-Lineage 1	WNV-Lineage 2	-
8	Zika virus	RNA	-	-	ZIKV
9	Dengue virus	RNA	DENV T1-4	-	-
10	*Plasmodium* species ID	DNA	*P. falciparum*	*P. ovm* [1]	-
11	*Plasmodium* infective stage	RNA	-	-	*Pf* infective stage

[1] *P. ovm*: *P. ovale/vivax/malariae*, i.e., the primer set detects all three. The symbol '-' means that there were no assay targets expected to be detected in the particular reaction chambers and detection channels.

2.5. VectorDisk Workflow

For the Vector Study, upon thawing, the mosquito lysate was diluted 1:200 with ultra-pure H_2O (RNase/DNase-free). Then, 180 μL of the diluted lysate was inserted into the VectorDisk inlet

(Figure 3A). In the automated process in the LabDisk, 160 µL of diluted lysate was transferred radially inwards (B) to the mixing chamber with the lyopellet (C), using temperature change-rate (TCR) valving [42] and centrifugo-dynamic inward pumping [43]. A dedicated microfluidic protocol was used in chamber (C) for the rigorous and homogeneous rehydration and mixing of the lyopellet, employing temperature change-rate actuated bubble-mixing [44,45]. The mixture was then transferred to the metering fingers (D), and a volume of 10 µL per finger was aliquoted using centrifugo-pneumatic aliquoting [46]. The valving and mixing operations were supported by a dedicated chamber for generating air pressure (E). Then the RT-PCR took place in the reaction chambers (F) where the primers/probes had been dry-stored. The protocol was: 44 °C for 10 min (reverse transcription), 97 °C for 2 min (initial denaturation for the PCR), and 40 cycles of 97 °C for 5 s and 60 °C for 4 s. The structure (F) serves to generate air pressure for valving and mixing operations. The structure (G) serves as a stiffening structure in order to prevent bending of the foil disk.

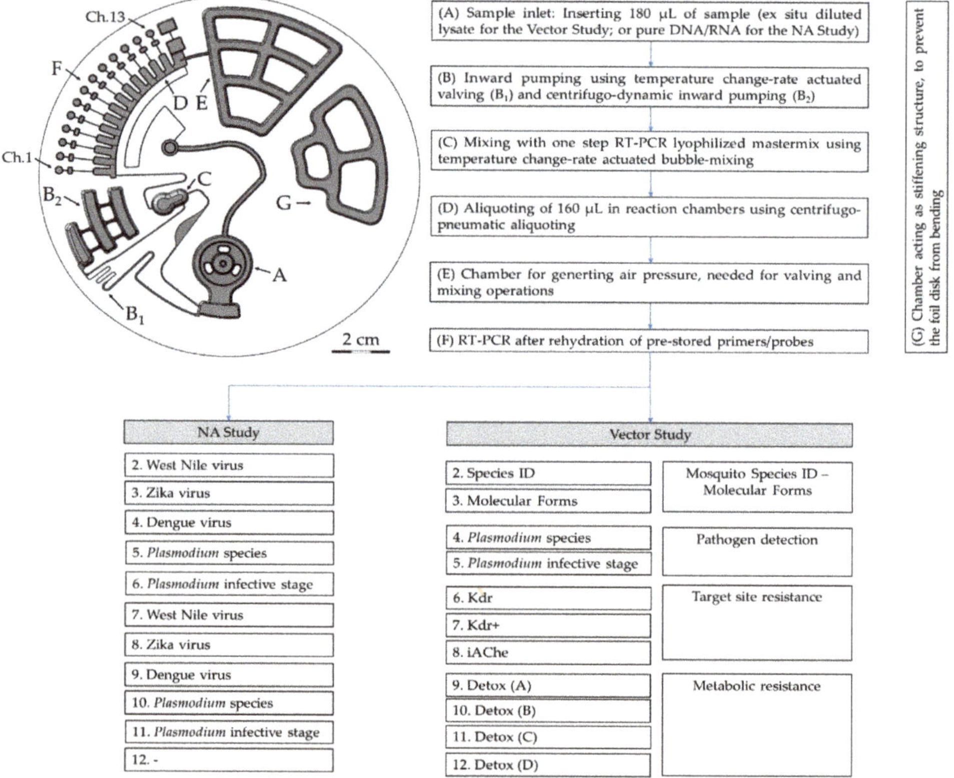

Figure 3. The VectorDisk design where the main unit operations are indicated (**A**–**F**). The structure G has the role of stiffening in order to prevent the foil disk from bending. The design was the same for both the Vector- and the NA-Studies. What differed was the type of inserted sample, and the assay panel. 'Ch.1, 2, ... ' refers to the VectorDisk reaction chambers.

For the NA Study, a mixture of 162 µL ultra-pure H_2O and 18 µL of purified NA sample was inserted into the inlet. Then, the same fluidic protocol was used as for the Vector Study. At the end, the PCR protocol was started: 50 °C for 15 min (reverse transcription), 95 °C for 3 min (initial denaturation for the PCR), and 40 or 45 cycles of 95 °C for 3 s and 60 °C for 30 s.

For both studies, the thermocycling protocol and the microfluidic processing were accomplished by means of a dedicated compact LabDisk Player instrument (Figure 4, QIAGEN Lake Constance, Stockach,

Germany). The detailed microfluidic protocol is described in Supplementary Table S3. Rotational frequency control and air-heating enabled centrifugo-pneumatic [43,46] and thermo-pneumatic microfluidic unit operations [42,44,47], and allowed for a time of 140–180 min from sample inlet to final result (depending on the Vector or NA Study). PCR raw data acquired with the LabDisk Player were converted and analyzed using the RotorGene Software (QIAGEN, Hilden, Germany).

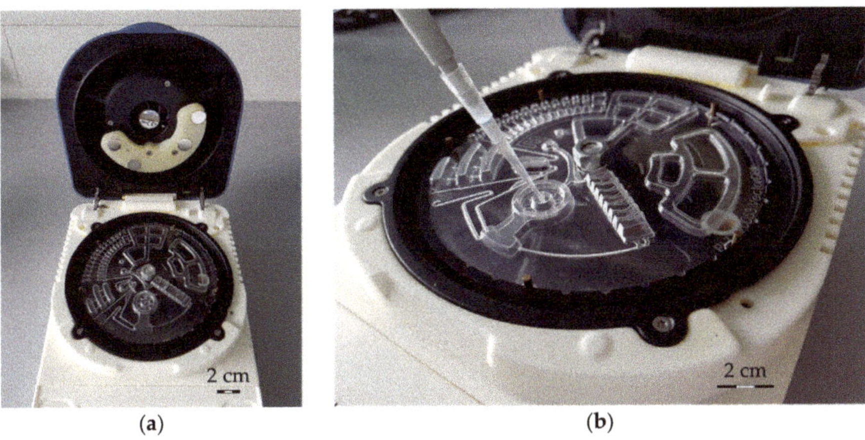

(a) (b)

Figure 4. (**a**) The LabDisk Player instrument used in the two studies. (**b**) Pipetting the sample (diluted lysate or purified nucleic acid (NA)) into the VectorDisk.

3. Results and Discussion

3.1. The Vector Study Results

In the Vector Study, we aimed to demonstrate the simultaneous detection of several DNA and RNA targets in triplex configuration. We tested 18 disks using diluted mosquito lysates as samples, and the arrangement of the assays (primers/probes per reaction chamber) is shown in Table 2.

This work focuses on the semi-quantitative screening of pooled specimens. The usefulness of the initial semi-quantitative screening of pooled mosquito specimens (detection/no detection of a mutation or pathogen) is supported by the two following paradigms. It is not uncommon to screen large numbers of individual mosquitoes and eventually find out that the whole population is wild type for a particular mutation (0% mutant allele frequency) or fixed for another one (100% mutant allele frequency). Furthermore, there are instances where the pathogen (*Plasmodium* or arbovirus) may not be present at all (low transmission settings) in several study sites, and these can be ruled out from a further, more detailed investigation.

The main goal of a semi-quantitative screening is to distinguish clearly between positive and negative results. This can be achieved by considering the gap between the maximum number of PCR cycles and the highest observed Ct value of a positive result. A maximum number of 40 or 45 PCR cycles (depending on the study) was fixed, according to authors' previous work taken as reference [34]. All targets which were not detected until that cycle number were considered as negative test results. The highest Ct values of positive results were in the range of 33 (Supplementary Table S4). The difference between that number and the maximum number of PCR cycles means that the concentration between the 'weakest' positive and the negative is at least two orders of magnitude. Therefore, we can clearly discriminate between positive and negative values. The RPS7 assay, targeting a house-keeping gene, can be considered as a positive amplification control, as it is expected to be positive in one of the three fluorescent wavelengths (FAM) and in four reaction chambers. To compare this with the assays validated in the preceding assay development work, the authors Mavridis et al. [34] had also used

40 cycles for their PCR (using a ViiA 7 Real-Time PCR System, Applied Biosystems, Waltham, MA, USA), and their highest Ct value of positive results was of the order of 34.

In Table 3 we present an overview of the true positive (TP), false negative (FN), true negative (TN), and false positive (FP) rates for each assay and for each mosquito pool, as well as in total for the Vector Study. The classification was done according to the expected outcomes based on previous work with these laboratory mosquitoes [29,31,34]. Some of the included assays were expected to be negative for the specific laboratory-reared mosquitoes (TN, also included in Table 3). In case a positive-expected assay was not detected, it was attributed to be false negative (FN). No FN results were found for the DNA assays. Notably, we also did not observe FP results in the Vector Study for any assay or in any of the five pools (Table 3), which demonstrates the high specificity of the disk-integrated assays. Some representative real-time PCR curves acquired by a VectorDisk are shown in Figure 5.

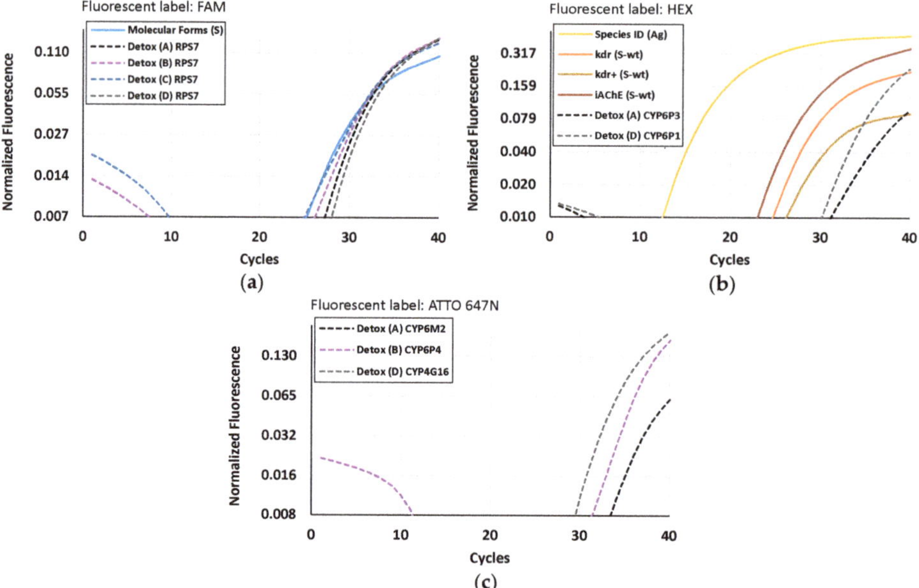

Figure 5. Indicative LabDisk real-time RT-PCR data acquired for the semi-quantitative analysis of mosquito samples in VectorDisk #D11, as one representative out of the 18 disks, from all three channels green (**a**), yellow (**b**), and red (**c**). DNA and RNA assays are shown in solid and dashed lines, respectively. The threshold was set to be the x-axis itself.

Another key output of this study was the detection of twelve RNA assays using a combination of geometric multiplexing (four reaction chambers: 9, 10, 11, 12) and color multiplexing (three wavelengths detecting in each chamber). Data from the assay development (previous work of the co-authors Mavridis et al. [34]) indicate that the triplex versus singleplex RT-PCR does not lead to loss of efficiency. The assays were successfully validated in terms of expression levels against standard two-step singleplex PCR assays and microarrays, using laboratory strains and field-caught samples. Further quantitative characterization of the efficiency, sensitivity, specificity, and reproducibility of the assays themselves is available in the co-authors' previous work [34]. In short, multiplex assays were efficient (reaction efficiencies = 95–109%), sensitive (covering a >10.0 Ct range up to Ct = 33.0 with R^2 values > 0.99), specific (TaqMan chemistry), and reproducible (CV = 4.5–12.1%).

Table 3. Overview of the true positive (TP), false negative (FN), true negative (TN), and false positive (FP) rates for each assay and for each mosquito pool, as well as in total for the Vector Study. The five different colors (Pool #1 … #5) refer to the five different pools of mosquitoes (Figure 1). The numbering in the first column indicates the VectorDisk reaction chamber where an assay is expected positive or negative. The grey horizontal shading in some assay rows (accompanied by the symbol '-') means that neither positive nor negative signal was expected, because the primer/probe mixtures stored in these chambers were expected to give signals in other detection channels.

No. of Lysate Pool		Pool #1				Pool #2				Pool #3				Pool #4				Pool #5				Overall			
Green detection channel		**TP**	**FN**	**TN**	**FP**	**TP**	**FN**	**TN**	**FP**	**TP**	**FN**	**TN**	**FP**	**TP**	**FN**	**TN**	**FP**	**TP**	**FN**	**TN**	**FP**	TP	FN	TN	FP
2	Species ID (Aq)			4/4	0/4			4/4	0/4			4/4	0/4			3/3	0/3			3/3	0/3			18/18	0/18
3	Molecular Forms (S)	4/4	0/4			4/4	0/4			4/4	0/4			3/3	0/3			3/3	0/3			18/18	0/18		
4	Kdr (Rw)			4/4	0/4			4/4	0/4			4/4	0/4			3/3	0/3			3/3	0/3			18/18	0/18
5	Kdr+ (R)			4/4	0/4			4/4	0/4			4/4	0/4			3/3	0/3			3/3	0/3			18/18	0/18
6	iAChe (R)			4/4	0/4			4/4	0/4			4/4	0/4			3/3	0/3			3/3	0/3			18/18	0/18
7	*P. falciparum*			4/4	0/4			4/4	0/4			4/4	0/4			3/3	0/3			3/3	0/3			18/18	0/18
8	-																								
9	Detox (A) RPS7	3/4	1/4			4/4	0/4			4/4	0/4			3/3	0/3			3/3	0/3			17/18	1/18		
10	Detox (B) RPS7	4/4	0/4			4/4	0/4			4/4	0/4			2/3	1/3			3/3	0/3			17/18	1/18		
11	Detox (C) RPS7	4/4	0/4			3/4	1/4			4/4	0/4			2/3	1/3			3/3	0/3			16/18	2/18		
12	Detox (D) RPS7	4/4	0/4			4/4	0/4			3/4	1/4			3/3	0/3			3/3	0/3			17/18	1/18		
Yellow detection channel		**TP**	**FN**	**TN**	**FP**	**TP**	**FN**	**TN**	**FP**	**TP**	**FN**	**TN**	**FP**	**TP**	**FN**	**TN**	**FP**	**TP**	**FN**	**TN**	**FP**	TP	FN	TN	FP
2	Species ID (Ag)	4/4	0/4			4/4	0/4			4/4	0/4			3/3	0/3			3/3	0/3			18/18	0/18		
3	Molecular Forms (M)			4/4	0/4			4/4	0/4			4/4	0/4			3/3	0/3			3/3	0/3			18/18	0/18
4	kdr (S-wt)	4/4	0/4			4/4	0/4			4/4	0/4			3/3	0/3			3/3	0/3			18/18	0/18		
5	kdr+ (S-wt)	4/4	0/4			4/4	0/4			4/4	0/4			3/3	0/3			3/3	0/3			18/18	0/18		
6	iAChE (S-wt)	4/4	0/4			4/4	0/4			4/4	0/4			3/3	0/3			3/3	0/3			18/18	0/18		
7	*Plasmodium ovm*			4/4	0/4			4/4	0/4			4/4	0/4			3/3	0/3			3/3	0/3			18/18	0/18
8	Infective stage (PLP1)			4/4	0/4			4/4	0/4			4/4	0/4			3/3	0/3			3/3	0/3			18/18	0/18
9	Detox (A) CYP6P3	3/4	1/4			4/4	0/4			3/4	1/4			3/3	0/3			3/3	0/3			16/18	2/18		
10	Detox (B) CYP9K1	1/4	3/4			2/4	2/4			2/4	2/4			2/3	1/3			1/3	2/3			8/18	10/18		
11	Detox (C) CYP6Z1	4/4	0/4			2/4	2/4			3/4	1/4			1/3	2/3			3/3	0/3			13/18	5/18		
12	Detox (D) CYP6P1	3/4	1/4			1/4	3/4			3/4	1/4			3/3	0/3			1/3	2/3			11/18	7/18		
Red detection channel		**TP**	**FN**	**TN**	**FP**	**TP**	**FN**	**TN**	**FP**	**TP**	**FN**	**TN**	**FP**	**TP**	**FN**	**TN**	**FP**	**TP**	**FN**	**TN**	**FP**	TP	FN	TN	FP
2	Species ID (Aa)			4/4	0/4			4/4	0/4			4/4	0/4			3/3	0/3			3/3	0/3			18/18	0/18
3	-																								
4	Kdr (Re)			4/4	0/4			4/4	0/4			4/4	0/4			3/3	0/3			3/3	0/3			18/18	0/18
5	-																								
6	-																								
7	-																								
8	-																								
9	Detox (A) CYP6M2	1/4	3/4			4/4	0/4			2/4	2/4			0/3	3/3			3/3	0/3			10/18	8/18		
10	Detox (B) CYP6P4	0/4	4/4			0/4	4/4			1/4	3/4			0/3	3/3			1/3	2/3			2/18	16/18		
11	Detox (C) GSTE2	4/4	0/4			2/4	2/4			3/4	1/4			0/3	3/3			2/3	1/3			11/18	7/18		
12	Detox (D) CYP4G16	3/4	1/4			2/4	2/4			3/4	1/4			3/3	0/3			0/3	3/3			11/18	7/18		

Contrary to the DNA assays, FN results were found among the RNA assays and, mostly, in the red channel. In particular, some of the Detox assays (e.g., Detox (B) K1 in yellow channel, Detox (A) CYP6M2 and Detox (B) CYP6P4 in red channel) exhibited a higher number of FN. The higher FN rate of the RNA assays may be attributed to the fact that RNA amplification includes one additional step—the RNA reverse transcription—which is more prone to inhibitors than the DNA amplification [48]. One more factor that may account for this higher FN rate in the case of RNA assays could possibly be the adsorption of reverse transcriptase to the disk surface. Given the fact that this enzyme is integrated into the lyophilized pellet that is stored in chamber C (Figure 3), undergoes rehydration and mixing, and the resultant liquid is then distributed to the reaction chambers (E), this appears to be an explanation. The above options are recorded as candidate investigation/optimization steps for the integration of the RNA-based insecticide resistance assays.

For an assessment of the intra-pool reproducibility of the results, we conducted a basic statistical analysis for each pool individually, which is summarized in Supplementary Table S4. Furthermore, and in order to have an indication of the inter-pool variation, we performed a weighted averaging (grey-colored 'Overall' columns of Supplementary Table S4), because for each specific assay we did not have the same number of data points (due to some FN cases). In an inter-pool comparison, the calculated CV of each assay was below 8% (the CV of the DNA assays ranging from 1.8% to 4.1%, while the CV of the RNA assays ranging from 3.2% to 7.9%). The CVs of individual assays in five intra-pool comparisons were also of this range. The main factors that may drive the inter-pool variation are the manual steps in disk preparation (pipetting of primers/probes, drying in the oven), and the manual preparation of mosquito samples, including the crude lysis. Therefore, the low variability—even between pools—can be attributed to the automated aliquoting into preloaded reaction cavities and is a strong indicator of the robust and reproducible performance of the VectorDisk. Regarding the intra-disk microfluidic reproducibility (i.e., among the reaction chambers), one could also make an assessment from the very comparable Ct values of the assay RPS7, targeting a house-keeping gene, which acts as normalizer, and which is detected in four chambers (9, 10, 11, and 12) in the green detection channel (Supplementary Table S4).

Overall, the Vector Study demonstrated the compatibility of the LabDisk platform with the crude lysis-direct amplification process. Results are in line with past work on assay development by the co-authors Mavridis et al. [34], who have shown that the crude lysis-direct amplification option performed very closely to a typical magnetic bead-based bind-wash-elute extraction and purification followed by PCR (comparison performed by means of the resulting Ct values). Furthermore, previous work on microfluidic integration by the authors has demonstrated the possibility of homogenizing biological matrices (e.g., human saliva [49], bacteria [50], cells [51], swabs [52]) using disk-integrated magnets that are actuated by external (still on LabDisk Player) magnets [53]. Therefore, upon suitable adaptation of the VectorDisk microfluidic design, the lysis and homogenization of mosquito pools could be translated from an ex situ manual (currently) process to an in situ automated (on disk) process, and be followed by in situ dilution and amplification towards a truly fully-automated vector-to-result analysis. This could offer an even higher degree of automation, and result in less overall analysis time, costs, and potential errors than performing manual mosquito processing and multiple PCR assays.

3.2. The Nucleic Acid Study Results

In the NA Study, the authors aimed to explore the applicability of the VectorDisk in the field of arboviral vectors in terms of the performance and reproducibility of the corresponding assays integrated in the disk. As no arboviral vectors were available, the experiments were done using nucleic acids (WNV-Lineage 1 and ZIKV). Furthermore, given that the mosquitoes used in the Vector Study were laboratory-reared and did not contain any malaria parasites, purified DNA and RNA from infected mosquitoes were used in the NA Study (sample 'PF-INF'). The purification process is described in previous work [29]. Nine disks were used, and the arrangement of the assays (primers/probes per reaction chamber) is shown in Table 2. From the results, we extracted true/false positivity/negativity rates, for each assay and in total, which are shown in Table 4. Each disk with a viral target was expected to give TP signals from two chambers where the corresponding primers and probes were pre-stored, and TN signals from all other eight chambers. Each disk with the sample mixture 'PF-INF' was expected to give TP signals from four chambers (#5 and 10 for DNA, and #6 and 11 for RNA detection), and TN signals from all other six chambers. This was indeed the case, and no FPs were observed, which indicates an excellent performance of specificity of the disk-integrated arboviral assays. Some representative real-time PCR curves are shown in Figure 6.

Table 4. Overview of the true positive (TP), false negative (FN), true negative (TN), and false positive (FP) rates for each assay and for each sample, as well as in total for the NA study. The numbering in the first column indicates the VectorDisk reaction chamber where an assay is expected positive or negative. The grey horizontal shading in some assay rows (accompanied by the symbol '-') means that neither positive nor negative signal was expected, because the primer/probe mixtures stored in these chambers were expected to give signals in other detection channels. The coloring of the columns 'WNV-Lineage 1', 'ZIKV', and 'PF-INF' corresponds to the coloring of Figure 2. 'PF-INF' represents the mixture of *P. falciparum* DNA + *P. falciparum* infective stage RNA.

	Sample	WNV-Lineage 1				ZIKV				PF-INF				Overall			
	Green detection channel	**TP**	**FN**	**TN**	**FP**	**TP**	**FN**	**TN**	**FP**	**TP**	**FN**	**TN**	**FP**	**TP**	**FN**	**TN**	**FP**
2	WNV-Lineage 1	3/3	0/3					3/3	0/3			3/3	0/3	3/3	0/3	6/6	0/6
3	-																
4	DENV T1-4			3/3	0/3			3/3	0/3			3/3	0/3			9/9	0/9
5	*Plasmodium falciparum*			3/3	0/3			3/3	0/3	3/3	0/3			3/3	0/3	6/6	0/6
6	-																
7	WNV-Lineage 1	3/3	0/3					3/3	0/3			3/3	0/3	3/3	0/3	6/6	0/6
8	-																
9	DENV T1-4			3/3	0/3			3/3	0/3			3/3	0/3			9/9	0/9
10	*Plasmodium falciparum*			3/3	0/3			3/3	0/3	3/3	0/3			3/3	0/3	6/6	0/6
11	-																
	Yellow detection channel	**TP**	**FN**	**TN**	**FP**	**TP**	**FN**	**TN**	**FP**	**TP**	**FN**	**TN**	**FP**	**TP**	**FN**	**TN**	**FP**
2	WNV-Lineage 2			3/3	0/3			3/3	0/3			3/3	0/3			9/9	0/9
3	-																
4	-																
5	*Plasmodium ovm* [1]			3/3	0/3			3/3	0/3			3/3	0/3			9/9	0/9
6	-																
7	WNV-Lineage 2			3/3	0/3			3/3	0/3			3/3	0/3			9/9	0/9
8	-																
9	-																
10	*Plasmodium ovm* [1]			3/3	0/3			3/3	0/3			3/3	0/3			9/9	0/9
11	-																
	Red detection channel	**TP**	**FN**	**TN**	**FP**	**TP**	**FN**	**TN**	**FP**	**TP**	**FN**	**TN**	**FP**	**TP**	**FN**	**TN**	**FP**
2	-																
3	ZIKV			3/3	0/3	3/3	0/3					3/3	0/3	3/3	0/3	6/6	0/6
4	-																
5	-																
6	*Pf* infective stage			3/3	0/3			3/3	0/3	3/3	0/3			3/3	0/3	6/6	0/6
7	-																
8	ZIKV			3/3	0/3	3/3	0/3					3/3	0/3	3/3	0/3	6/6	0/6
9	-																
10	-																
11	*Pf* infective stage			3/3	0/3			3/3	0/3	3/3	0/3			3/3	0/3	6/6	0/6

[1] *P. ovm*: *P. ovale/vivax/malariae*, i.e., the primer set detects all three.

The disks also included primer/probes for detection of dengue virus (DENV T1-4), WNV-Lineage 2, and *P. ovale/vivax/malariae* (Table 2), but since the corresponding targets were not present in any of the samples, only TN or FP could be expected from these assays. Therefore, these primers/probes were used for cross-specificity purposes (tested against the other samples) and notably, only TN results were extracted. Furthermore, after analyzing the data for a reproducibility assessment, it is noteworthy that all assays in the VectorDisk performed with an inter-disk CV < 6% (Supplementary Table S5).

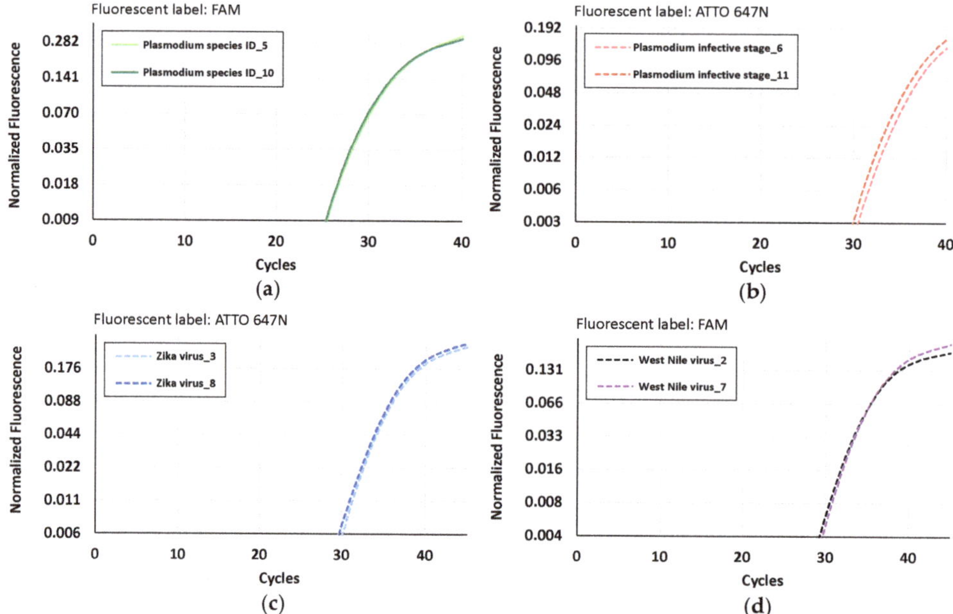

Figure 6. Representative LabDisk real-time PCR, acquired from the analysis of purified RNA for the various targets of the NA Study. (**a**) *Pf* species ID (DNA); (**b**) *Pf* infective stage (RNA); (**c**) ZIKV (RNA); (**d**) WNV-Lineage 1 (RNA). Each VectorDisk graph has two curves because each assay was included in two reaction chambers per disk (indicated in the figure legends).

Overall, the NA Study was performed by using the same disk design and fluidic operations, as well as the same lyopellet as the Vector Study, by only exchanging the primer/probes in the reaction chambers, and by slightly modifying the PCR protocol. This easy adaptability to the needs of the analysis is of utmost importance for an automated system, as it allows for a higher volume of manufacturing, resulting in lower costs (several different requests can be served by the same disk design). Furthermore, such a configuration allows a fast response to urgent needs—for instance, in the case of epidemics—because the primers/probes are de-coupled from the amplification reagents. Also important is the fact that the reagents have been developed and integrated in a way that they require no cold chain for transport, which is essential when the target settings are tropical regions. A comprehensive review of candidate analytical methods and platforms has been published by Vontas et al. [16]. The fact that the LabDisk platform has been shown to successfully detect tropical infections on human samples (malaria, dengue, chikungunya) [26,54,55], in combination with the current demonstration of compatibility with analysis of multiple vector-related assays, paves the way for a One Health approach of combined human/vector diagnostics [22].

4. Conclusions

This work included two proof-of-principle studies. The Vector Study demonstrated the capacity of the VectorDisk to simultaneously and semi-quantitatively detect DNA- and RNA-specific targets in multiplex assays from mosquito pool samples, which were crudely lysed, diluted, and then directly amplified in the disk. A combination of multiple microfluidic reaction chambers and detection wavelengths (geometric and color multiplexing) was successfully used for this scope. The NA Study, starting from purified nucleic acids, demonstrated the rapid adaptability of the platform by changing only the primer/probe panel and the capacity to detect arboviral vector and malaria-infected vector assays. In both studies, key targets relevant to species identification, infective stage, and insecticide resistance markers, were detected. Both studies indicated a high reproducibility in detection of the assays, with intra-and inter-pool CVs of the PCR Ct values being <10%. High specificity was also achieved with no FPs detected in either of the Vector or NA Studies.

Overall, the outcomes of our work give a promising perspective on the platform to be applied in fully automated vector diagnostics—a platform which, once tested and validated with field-collected disease vectors, will contribute molecular entomological data that is highly relevant to decision-making in vector control programs. Regarding this implementation, we foresee two further steps: (i) the in situ (on-disk) lysis and homogenization of mosquitoes that will enable fully integrated sample preparation, which can be advantageous for minimizing hands-on work; and (ii) the transition from semi-quantitative to quantitative analysis, which will offer more detailed information related to insecticide resistances, contributing to vector surveillance and control.

Supplementary Materials: The following are available online at http://www.mdpi.com/2227-9717/8/12/1677/s1, Table S1: Primers/probes sequences for the assays in the Vector Study, Table S2: Primers/probes sequences for the assays in the NA Study, Table S3: Microfluidic protocol to operate the VectorDisk, Table S4: Statistical analysis of the Ct values derived from five pools of mosquitoes tested with the VectorDisk during the Vector Study, Table S5: Statistical analysis of the Ct values derived from the VectorDisk results from the NA Study.

Author Contributions: Conceptualization, S.H., D.B., M.S., N.W., P.M., K.M. (Konstantinos Mavridis), J.V. and K.M. (Konstantinos Mitsakakis); methodology, S.H., D.B., M.S., B.J., M.R. and K.M. (Konstantinos Mitsakakis); software, E.M.A. and M.R.; validation, S.H., D.B., M.S. and E.M.A.; formal analysis, D.B.; investigation, S.H., D.B., M.S., J.L., E.M.A. and L.S.; resources, N.W., P.M., K.M. (Konstantinos Mavridis) and J.V.; data curation, S.H. and D.B.; writing—original draft preparation, D.B. and K.M. (Konstantinos Mitsakakis); writing—review and editing, S.H., D.B., B.J., F.v.S., N.W., P.M., K.M. (Konstantinos Mavridis) and K.M. (Konstantinos Mitsakakis); visualization, S.H., D.B. and K.M. (Konstantinos Mitsakakis); supervision, S.H., M.R., N.P., R.Z. and K.M. (Konstantinos Mitsakakis); project administration, S.H., P.M., K.M. (Konstantinos Mavridis), J.V. and K.M. (Konstantinos Mitsakakis); funding acquisition, P.M., J.V. and K.M. (Konstantinos Mitsakakis) All authors have read and agreed to the published version of the manuscript.

Funding: This project has received funding from the European Union's Horizon 2020 research and innovation programme under grant agreements No 688207 ('DMC-MALVEC' project) and 731060 ('Infravec2' project). The article processing charge was funded by the Baden-Wuerttemberg Ministry of Science, Research and Art and the University of Freiburg in the funding programme Open Access Publishing.

Acknowledgments: The authors would like to acknowledge the Hahn-Schickard Lab-on-a-Chip Foundry Service for LabDisk production as well as BEI Resources for providing the viral genomic RNA samples used in this study. The insectary at the Swiss Tropical and Public Health Institute provided specimens of the Kisumu colony that originates from an MRA-762 egg batch provided by BEI Resources.

Conflicts of Interest: The authors declare no conflict of interest. The funders had no role in the design of the study; in the collection, analyses, or interpretation of data; in the writing of the manuscript, or in the decision to publish the results.

References

1. WHO. *Global Vector Control Response 2017–2030*; World Health Organization (WHO): Geneva, Switzerland, 2017; Available online: https://www.who.int/vector-control/publications/global-control-response/en (accessed on 18 November 2020).
2. Müller, R.; Reuss, F.; Kendrovski, V.; Montag, D. Vector-borne diseases. In *Biodiversity and Health in the Face of Climate Change*; Marselle, M., Stadler, J., Korn, H., Irvine, K., Bonn, A., Eds.; Springer: Cham, Switzerland, 2019; pp. 67–90. [CrossRef]
3. Vorou, R. Zika virus, vectors, reservoirs, amplifying hosts, and their potential to spread worldwide: What we know and what we should investigate urgently. *Int. J. Infect. Dis.* **2016**, *48*, 85–90. [CrossRef]
4. Heyman, P.; Cochez, C.; Hofhuis, A.; van der Giessen, J.; Sprong, H.; Porter, S.R.; Losson, B.; Saegerman, C.; Donoso-Mantke, O.; Niedrig, M.; et al. A clear and present danger: Tick-borne diseases in Europe. *Expert Rev. Anti Infect. Ther.* **2010**, *8*, 33–50. [CrossRef]
5. Rochlin, I.; Ninivaggi, D.V.; Hutchinson, M.L.; Farajollahi, A. Climate change and range expansion of the Asian Tiger Mosquito (*Aedes albopictus*) in northeastern USA: Implications for public health practitioners. *PLoS ONE* **2013**, *8*, e60874. [CrossRef]

6. Medlock, J.M.; Hansford, K.M.; Schaffner, F.; Versteirt, V.; Hendrickx, G.; Zeller, H.; Van Bortel, W. A review of the invasive mosquitoes in Europe: Ecology, public health risks, and control options. *Vector Borne Zoonotic Dis.* **2012**, *12*, 435–447. [CrossRef]
7. Rezza, G.; Nicoletti, L.; Angelini, R.; Romi, R.; Finarelli, A.C.; Panning, M.; Cordioli, P.; Fortuna, C.; Boros, S.; Magurano, F.; et al. Infection with chikungunya virus in Italy: An outbreak in a temperate region. *Lancet* **2007**, *370*, 1840–1846. [CrossRef]
8. Venturi, G.; Di Luca, M.; Fortuna, C.; Remoli, M.E.; Riccardo, F.; Severini, F.; Toma, L.; Del Manso, M.; Benedetti, E.; Caporali, M.G.; et al. Detection of a chikungunya outbreak in Central Italy, August to September 2017. *Eurosurveillance* **2017**, *22*, 11–14. [CrossRef]
9. Giron, S.; Franke, F.; Decoppet, A.; Cadiou, B.; Travaglini, T.; Thirion, L.; Durand, G.; Jeannin, C.; L'Ambert, G.; Grard, G.; et al. Vector-borne transmission of Zika virus in Europe, southern France, August 2019. *Eurosurveillance* **2019**, *24*, 2–5. [CrossRef] [PubMed]
10. Kioulos, I.; Kampouraki, A.; Morou, E.; Skavdis, G.; Vontas, J. Insecticide resistance status in the major West Nile virus vector Culex pipiens from Greece. *Pest Manag. Sci.* **2014**, *70*, 623–627. [CrossRef] [PubMed]
11. Pervanidou, D.; Vakali, A.; Georgakopoulou, T.; Panagiotopoulos, T.; Patsoula, E.; Koliopoulos, G.; Politis, C.; Stamoulis, K.; Gavana, E.; Pappa, S.; et al. West Nile virus in humans, Greece, 2018: The largest seasonal number of cases, 9 years after its emergence in the country. *Eurosurveillance* **2020**, *25*, 15–27. [CrossRef]
12. Gossner, C.M.; Marrama, L.; Carson, M.; Allerberger, F.; Calistri, P.; Dilaveris, D.; Lecollinet, S.; Morgan, D.; Nowotny, N.; Paty, M.; et al. West Nile virus surveillance in Europe: Moving towards an integrated animal-human-vector approach. *Eurosurveillance* **2017**, *22*, 10–19. [CrossRef]
13. Bhatt, S.; Weiss, D.J.; Cameron, E.; Bisanzio, D.; Mappin, B.; Dalrymple, U.; Battle, K.; Moyes, C.L.; Henry, A.; Eckhoff, P.A.; et al. The effect of malaria control on *Plasmodium falciparum* in Africa between 2000 and 2015. *Nature* **2015**, *526*, 207–211. [CrossRef] [PubMed]
14. Ranson, H.; Lissenden, N. Insecticide resistance in African Anopheles Mosquitoes: A worsening situation that needs urgent action to maintain malaria control. *Trends Parasitol.* **2015**, *32*, 187–196. [CrossRef] [PubMed]
15. Hemingway, J.; Field, L.; Vontas, J. An overview of insecticide resistance. *Science* **2002**, *298*, 96–97. [CrossRef] [PubMed]
16. Vontas, J.; Mavridis, K. Vector population monitoring tools for insecticide resistance management: Myth or fact? *Pest. Biochem. Physiol.* **2019**, *162*, 54–60. [CrossRef] [PubMed]
17. Bass, C.; Williamson, M.S.; Field, L.M. Development of a multiplex real-time PCR assay for identification of members of the Anopheles gambiae species complex. *Acta Trop.* **2008**, *107*, 50–53. [CrossRef] [PubMed]
18. Bass, C.; Nikou, D.; Blagborough, A.M.; Vontas, J.; Sinden, R.E.; Williamson, M.S.; Field, L.M. PCR-based detection of Plasmodium in Anopheles mosquitoes: A comparison of a new high-throughput assay with existing methods. *Malar. J.* **2008**, *7*, 177. [CrossRef]
19. Bass, C.; Nikou, D.; Vontas, J.; Donnelly, M.J.; Williamson, M.S.; Field, L.M. The vector population monitoring tool (VPMT): High-throughput DNA-based diagnostics for the monitoring of mosquito vector populations. *Malar. Res. Treat.* **2010**, *2010*, 190434. [CrossRef]
20. Donnelly, M.J.; Isaacs, A.T.; Weetman, D. Identification, validation, and application of molecular diagnostics for insecticide resistance in malaria vectors. *Trends Parasitol.* **2016**, *32*, 197–206. [CrossRef]
21. Weetman, D.; Donnelly, M.J. Evolution of insecticide resistance diagnostics in malaria vectors. *Trans. R. Soc. Trop. Med. Hyg.* **2015**, *109*, 291–293. [CrossRef]
22. Mitsakakis, K.; Hin, S.; Muller, P.; Wipf, N.; Thomsen, E.; Coleman, M.; Zengerle, R.; Vontas, J.; Mavridis, K. Converging human and malaria vector diagnostics with data management towards an integrated holistic one health approach. *Int. J. Environ. Res. Public Health* **2018**, *15*, 259. [CrossRef]
23. WHO. *Microscopy*; World Health Organization: Geneva, Switzerland, 2017. Available online: www.who.int/malaria/areas/diagnosis/microscopy/en (accessed on 9 August 2020).
24. Vontas, J.; Mitsakakis, K.; Zengerle, R.; Yewhalaw, D.; Sikaala, C.H.; Etang, J.; Fallani, M.; Carman, B.; Muller, P.; Chouaibou, M.; et al. Automated innovative diagnostic, data management and communication tool, for improving malaria vector control in endemic settings. *Stud. Health Technol. Inform.* **2016**, *224*, 54–60. [PubMed]

25. Czilwik, G.; Messinger, T.; Strohmeier, O.; Wadle, S.; von Stetten, F.; Paust, N.; Roth, G.; Zengerle, R.; Saarinen, P.; Niittymaki, J.; et al. Rapid and fully automated bacterial pathogen detection on a centrifugal-microfluidic LabDisk using highly sensitive nested PCR with integrated sample preparation. *Lab Chip* **2015**, *15*, 3749–3759. [CrossRef] [PubMed]
26. Hin, S.; Lopez-Jimena, B.; Bakheit, M.; Klein, V.; Stack, S.; Fall, C.; Sall, A.; Enan, K.; Frischmann, S.; Gillies, L.; et al. The FeverDisk: Multiplex detection of fever-causing pathogens for rapid diagnosis of tropical diseases. In Proceedings of the 21st International Conference on Miniaturized Systems for Chemistry and Life Sciences, µTAS 2017, Savannah, GA, USA, 22–26 October 2017; pp. 7–8.
27. Rombach, M.; Hin, S.; Specht, M.; Johannsen, B.; Lüddecke, J.; Paust, N.; Zengerle, R.; Roux, L.; Sutcliffe, T.; Pecham, J.R.; et al. RespiDisk: A point-of-care platform for fully automated detection of respiratory tract infection pathogens in clinical samples. *Analyst* **2020**, *145*, 7040–7047. [CrossRef] [PubMed]
28. Vernick, K.D. Infravec2: Expanding researcher access to insect vector tools and resources. *Pathog. Glob. Health* **2017**, *111*, 217–218. [CrossRef] [PubMed]
29. Kefi, M.; Mavridis, K.; Simoes, M.L.; Dimopoulos, G.; Siden-Kiamos, I.; Vontas, J. New rapid one-step PCR diagnostic assay for Plasmodium falciparum infective mosquitoes. *Sci. Rep.* **2018**, *8*, 1462. [CrossRef] [PubMed]
30. Bass, C.; Nikou, D.; Donnelly, M.J.; Williamson, M.S.; Ranson, H.; Ball, A.; Vontas, J.; Field, L.M. Detection of knockdown resistance (kdr) mutations in Anopheles gambiae: A comparison of two new high-throughput assays with existing methods. *Malar. J.* **2007**, *6*, 111. [CrossRef]
31. Mavridis, K.; Wipf, N.; Müller, P.; Traore, M.M.; Muller, G.; Vontas, J. Detection and monitoring of insecticide resistance mutations in *Anopheles gambiae*: Individual vs pooled specimens. *Genes* **2018**, *9*, 479. [CrossRef]
32. Jones, C.M.; Liyanapathirana, M.; Agossa, F.R.; Weetman, D.; Ranson, H.; Donnelly, M.J.; Wilding, C.S. Footprints of positive selection associated with a mutation (N1575Y) in the voltage-gated sodium channel of *Anopheles gambiae*. *Proc. Natl. Acad. Sci. USA* **2012**, *109*, 6614–6619. [CrossRef]
33. Bass, C.; Nikou, D.; Vontas, J.; Williamson, M.S.; Field, L.M. Development of high-throughput real-time PCR assays for the identification of insensitive acetylcholinesterase (ace-1R) in Anopheles gambiae. *Pestic. Biochem. Physiol.* **2010**, *96*, 80–85. [CrossRef]
34. Mavridis, K.; Wipf, N.; Medves, S.; Erquiaga, I.; Müller, P.; Vontas, J. Rapid multiplex gene expression assays for monitoring metabolic resistance in the major malaria vector *Anopheles gambiae*. *Parasites Vectors* **2019**, *12*, 9. [CrossRef]
35. Del Amo, J.; Sotelo, E.; Fernandez-Pinero, J.; Gallardo, C.; Llorente, F.; Aguero, M.; Jimenez-Clavero, M.A. A novel quantitative multiplex real-time RT-PCR for the simultaneous detection and differentiation of West Nile virus lineages 1 and 2, and of Usutu virus. *J. Virol. Methods* **2013**, *2*, 321–327. [CrossRef] [PubMed]
36. Chan, J.F.W.; Yip, C.C.Y.; Tee, K.M.; Zhu, Z.; Tsang, J.O.L.; Chik, K.K.H.; Tsang, T.G.W.; Chan, C.C.S.; Poon, V.K.M.; Sridhar, S.; et al. Improved detection of Zika virus RNA in human and animal specimens by a novel, highly sensitive and specific real-time RT-PCR assay targeting the 5'-untranslated region of Zika virus. *Trop. Med. Int. Health* **2017**, *22*, 594–603. [CrossRef] [PubMed]
37. Hahn-Schickard Lab-on-a-Chip Foundry. Available online: https://www.hahn-schickard.de/en/production/lab-on-a-chip-foundry (accessed on 18 November 2020).
38. Focke, M.; Stumpf, F.; Faltin, B.; Reith, P.; Bamarni, D.; Wadle, S.; Müller, C.; Reinecke, H.; Schrenzel, J.; Francois, P. Microstructuring of polymer films for sensitive genotyping by real-time PCR on a centrifugal microfluidic platform. *Lab Chip* **2010**, *10*, 2519–2526. [CrossRef] [PubMed]
39. Focke, M.; Kosse, D.; Al-Bamerni, D.; Lutz, S.; Müller, C.; Reinecke, H.; Zengerle, R.; von Stetten, F. Microthermoforming of microfluidic substrates by soft lithography (µTSL): Optimization using design of experiments. *J. Micromech. Microeng.* **2011**, *21*, 115002. [CrossRef]
40. Simmons, M.; Myers, T.; Guevara, C.; Jungkind, D.; Williams, M.; Houng, H.S. Development and validation of a quantitative, one-step, multiplex, real-time reverse transcriptase PCR assay for detection of dengue and Chikungunya viruses. *J. Clin. Microbiol.* **2016**, *54*, 1766–1773. [CrossRef]
41. Kosse, D.; Buselmeier, D.; Müller, C.; Zengerle, Z.; von Stetten, F. Gas pressure assisted thermal bonding of film-based Lab-on-a-Chip cartridges. In Proceedings of the Mikrosystemtechnik-Kongress, Darmstadt, Germany, 10–12 October 2011; pp. 579–582.

42. Keller, M.; Czilwik, G.; Schott, J.; Schwarz, I.; Dormanns, K.; von Stetten, F.; Zengerle, R.; Paust, N. Robust temperature change rate actuated valving and switching for highly integrated centrifugal microfluidics. *Lab Chip* **2017**, *17*, 864–875. [CrossRef]
43. Zehnle, S.; Schwemmer, F.; Roth, G.; von Stetten, F.; Zengerle, R.; Paust, N. Centrifugo-dynamic inward pumping of liquids on a centrifugal microfluidic platform. *Lab Chip* **2012**, *12*, 5142–5145. [CrossRef]
44. Hin, S.; Paust, N.; Keller, M.; Rombach, M.; Strohmeier, O.; Zengerle, R.; Mitsakakis, K. Temperature change rate actuated bubble mixing for homogeneous rehydration of dry pre-stored reagents in centrifugal microfluidics. *Lab Chip* **2018**, *18*, 362–370. [CrossRef]
45. Burger, S.; Schulz, M.; von Stetten, F.; Zengerle, R.; Paust, N. Rigorous buoyancy driven bubble mixing for centrifugal microfluidics. *Lab Chip* **2016**, *16*, 261–268. [CrossRef]
46. Mark, D.; Weber, P.; Lutz, S.; Focke, M.; Zengerle, R.; von Stetten, F. Aliquoting on the centrifugal microfluidic platform based on centrifugo-pneumatic valves. *Microfluid. Nanofluid.* **2011**, *10*, 1279–1288. [CrossRef]
47. Strohmeier, O.; Keller, M.; Schwemmer, F.; Zehnle, S.; Mark, D.; von Stetten, F.; Zengerle, R.; Paust, N. Centrifugal microfluidic platforms: Advanced unit operations and applications. *Chem. Soc. Rev.* **2015**, *44*, 6187–6229. [CrossRef] [PubMed]
48. Fleige, S.; Pfaffl, M.W. RNA integrity and the effect on the real-time qRT-PCR performance. *Mol. Asp. Med.* **2006**, *27*, 126–139. [CrossRef] [PubMed]
49. Johannsen, B.; Müller, L.; Baumgartner, D.; Karkossa, L.; Fruh, S.M.; Bostanci, N.; Karpisek, M.; Zengerle, R.; Paust, N.; Mitsakakis, K. Automated pre-analytic processing of whole saliva using magnet-beating for point-of-care protein biomarker analysis. *Micromachines* **2019**, *10*, 833. [CrossRef] [PubMed]
50. Kim, J.; Jang, S.H.; Jia, G.Y.; Zoval, J.V.; Da Silva, N.A.; Madou, M.J. Cell lysis on a microfluidic CD (compact disc). *Lab Chip* **2004**, *4*, 516–522. [CrossRef]
51. Kido, H.; Micic, M.; Smith, D.; Zoval, J.; Norton, J.; Madou, M. A novel, compact disk-like centrifugal microfluidics system for cell lysis and sample homogenization. *Colloid Surf. B Biointerfaces* **2007**, *58*, 44–51. [CrossRef]
52. Siegrist, J.; Gorkin, R.; Bastien, M.; Stewart, G.; Peytavi, R.; Kido, H.; Bergeron, M.; Madou, M. Validation of a centrifugal microfluidic sample lysis and homogenization platform for nucleic acid extraction with clinical samples. *Lab Chip* **2010**, *10*, 363–371. [CrossRef]
53. Hin, S.; Paust, N.; Rombach, M.; Lueddecke, J.; Specht, M.; Zengerle, R.; Mitsakakis, K. Minimizing ethanol carry-over in centrifugal microfluidic nucleic acid extraction by advanced bead handling and management of diffusive mass transfer. In Proceedings of the 20th International Conference on Solid-State Sensors, Actuators and Microsystems & Eurosensors XXXIII, Berlin, Germany, 23–27 June 2019; pp. 130–133. [CrossRef]
54. Lopez-Jimena, B.; Bekaert, M.; Bakheit, M.; Frischmann, S.; Patel, P.; Simon-Loriere, E.; Lambrechts, L.; Duong, V.; Dussart, P.; Harold, G.; et al. Development and validation of four one-step real-time RT-LAMP assays for specific detection of each dengue virus serotype. *PLoS Negl. Trop. Dis.* **2018**, *12*, e0006381. [CrossRef]
55. Lopez-Jimena, B.; Wehner, S.; Harold, G.; Bakheit, M.; Frischmann, S.; Bekaert, M.; Faye, O.; Sall, A.A.; Weidmann, M. Development of a single-tube one-step RT-LAMP assay to detect the Chikungunya virus genome. *PLoS Negl. Trop. Dis.* **2018**, *12*, e0006448. [CrossRef]

Article

Microfluidic Nano-Scale qPCR Enables Ultra-Sensitive and Quantitative Detection of SARS-CoV-2

Xin Xie [1], Tamara Gjorgjieva [2,3], Zaynoun Attieh [1], Mame Massar Dieng [2], Marc Arnoux [4], Mostafa Khair [4], Yasmine Moussa [1], Fatima Al Jallaf [2,3], Nabil Rahiman [1], Christopher A. Jackson [5], Lobna El Messery [6], Khristine Pamplona [6], Zyrone Victoria [6], Mohammed Zafar [6], Raghib Ali [3], Fabio Piano [1,5], Kristin C. Gunsalus [1,3,5,*] and Youssef Idaghdour [2,3,*]

1 Center for Genomics and Systems Biology, New York University Abu Dhabi, P.O. Box 129188, Abu Dhabi 51133, UAE; xx12@nyu.edu (X.X.); za2061@nyu.edu (Z.A.); yasminemoussa@nyu.edu (Y.M.); nabil.rahiman@nyu.edu (N.R.); fp1@nyu.edu (F.P.)
2 Program in Biology, Division of Science, New York University Abu Dhabi, P.O. Box 129188, Abu Dhabi 51133, UAE; tg1407@nyu.edu (T.G.); mmd14@nyu.edu (M.M.D.); fa69@nyu.edu (F.A.J.)
3 Public Health Research Center, New York University Abu Dhabi, P.O. Box 129188, Abu Dhabi 51133, UAE; ra107@nyu.edu
4 Core Technology Platforms, New York University Abu Dhabi, P.O. Box 129188, Abu Dhabi 51133, UAE; mga5@nyu.edu (M.A.); mrk6@nyu.edu (M.K.)
5 Department of Biology and Center for Genomics and Systems Biology, New York University, New York, NY 10003, USA; cj59@nyu.edu
6 Proficiency Healthcare Diagnostics, Electra Street, Abu Dhabi 51133, UAE; dr.l.elmessery@proficiencylab.org (L.E.M.); khristinep@proficiencylab.org (K.P.); zyronev@proficiencylab.org (Z.V.); zafar.m@proficiencylab.org (M.Z.)
* Correspondence: kcg1@nyu.edu (K.C.G.); youssef.idaghdour@nyu.edu (Y.I.)

Received: 30 September 2020; Accepted: 25 October 2020; Published: 9 November 2020

Abstract: A major challenge in controlling the COVID-19 pandemic is the high false-negative rate of the commonly used RT-PCR methods for SARS-CoV-2 detection in clinical samples. Accurate detection is particularly challenging in samples with low viral loads that are below the limit of detection (LoD) of standard one- or two-step RT-PCR methods. In this study, we implemented a three-step approach for SARS-CoV-2 detection and quantification that employs reverse transcription, targeted cDNA preamplification, and nano-scale qPCR based on a commercially available microfluidic chip. Using SARS-CoV-2 synthetic RNA and plasmid controls, we demonstrate that the addition of a preamplification step enhances the LoD of this microfluidic RT-qPCR by 1000-fold, enabling detection below 1 copy/μL. We applied this method to analyze 182 clinical NP swab samples previously diagnosed using a standard RT-qPCR protocol (91 positive, 91 negative) and demonstrate reproducible and quantitative detection of SARS-CoV-2 over five orders of magnitude (<1 to 10^6 viral copies/μL). Crucially, we detect SARS-CoV-2 with relatively low viral load estimates (<1 to 40 viral copies/μL) in 17 samples with negative clinical diagnosis, indicating a potential false-negative rate of 18.7% by clinical diagnostic procedures. In summary, this three-step nano-scale RT-qPCR method can robustly detect SARS-CoV-2 in samples with relatively low viral loads (<1 viral copy/μL) and has the potential to reduce the false-negative rate of standard RT-PCR-based diagnostic tests for SARS-CoV-2 and other viral infections.

Keywords: SARS-CoV-2; COVID-19; microfluidics; nano-qPCR; ultra-sensitive; viral RNA; viral load; detection

1. Introduction

The most widely used method for the detection of SARS-CoV-2 (the infectious agent that causes COVID-19) in nasopharyngeal (NP) swab samples is Reverse Transcription Polymerase Chain Reaction (RT-PCR) [1–3]. Inaccurate test results from RT-PCR have been widely reported, with estimated false-negative rates of 10–30% among different implementations of this method [4–7]. Such high false-negative rates pose a significant challenge to controlling the spread of infection, and are further exacerbated by poor sample quality or low viral loads that are below the detection limit of standard RT-PCR methods [8,9]. This, combined with both the cost and scarcity of reagents [10–12], has hampered global scale-up of RT-PCR testing to levels that would be required to adequately monitor communities for COVID-19.

An additional challenge to controlling the spread of COVID-19 is the role of asymptomatic transmission [13–16]. Different estimates suggest that 40 to 80% of infected individuals are either pre-symptomatic, asymptomatic, or mildly symptomatic [17–20]. Early detection of infection in these individuals is crucial for disease control, which is why many countries and communities have started implementing active screening programs that extend COVID-19 testing to asymptomatic individuals. However, asymptomatic carriers sometimes carry very low viral loads [21] that may not be detected by a standard RT-PCR test [22]. Therefore, the development of more sensitive detection methods that can detect low viral loads is crucial.

Most commercial kits for COVID-19 testing utilize either a one-step RT-PCR approach, which combines the RT and qPCR reactions, or a two-step approach in which RT and qPCR are performed sequentially. A target-specific preamplification step has been successfully incorporated in a number of studies to detect and analyze various types of samples with limited amount of genomic materials, including viruses in human samples [23] or in drinking water [24], circulating tumor DNA in blood [25], and even ancient DNA samples [26]. Therefore, in this study, we implemented a three-step approach involving sequential RT, targeted cDNA preamplification, and qPCR, using a commercially available microfluidics platform. Using this method, we demonstrate reliable ultra-sensitive detection of low SARS-CoV-2 viral loads in both standard positive controls and clinical NP swab samples, including samples previously diagnosed as negative by an accredited diagnostic lab. Overall, this microfluidic RT-PCR assay is a cost-effective strategy with the potential to reduce the false-negative rate of clinical diagnostic tests, and as such, could be a valuable tool in active screening programs aimed at the early detection of SARS-CoV-2.

2. Materials and Methods

We first implemented the three-step SARS-CoV-2 detection method using synthetic SARS-CoV-2 RNA and SARS-CoV-2 plasmids and determined the limit of detection, before validating this method in clinical nasopharyngeal swab samples.

2.1. Ethics Statement

This study was determined as exempt by the NYU Abu Dhabi (NYUAD) Institutional Review Board (HRPP-2020-48) as it involves clinical samples that have already been collected by a diagnostic lab for the primary purpose of SARS-CoV-2 testing, and have been de-identified before being transported to and analyzed at NYUAD.

2.2. Positive Controls

Two types of positive controls were used in this study. The Twist Synthetic SARS-CoV-2 RNA (102024, Twist Biosciences, San Francisco, CA, USA) consists of six non-overlapping ssRNA fragments with a coverage of greater than 99.9% of the viral genome. The SARS-CoV-2 plasmids (10006625, IDT, Leuven, Belgium) contain the complete DNA sequence of the SARS-CoV-2 nucleocapsid gene.

2.3. SARS-CoV-2 Detection (Synthetic RNA and Plasmid)

Two assays (primer/probe sets) were used for SARS-CoV-2 detection, per CDC recommendations: 2019-nCoV_N1 and 2019-nCoV_N2 (2019-nCoV CDC EUA Kit, 10006606, IDT). The human RNase P (RP) assay was used as a control for RNA extraction and RT-qPCR reactions. For both positive controls, 10-fold serial dilutions were prepared, with two replicates at each concentration. Each sample was analyzed using 9 replicates for N1, 9 replicates for N2, and 6 replicates for RP assays.

Manual extraction of synthetic SARS-CoV-2 RNA was performed using the ABIOpureTM Viral DNA/RNA Extraction Kit (M561VT50, Alliance Bio, Bothell, WA, USA) according to the manufacturer's instructions. The isolated synthetic RNA/SARS-CoV-2 plasmids were then used for reverse transcription (RT) and quantitative PCR (qPCR) using the Fluidigm Real-Time PCR for Viral RNA Detection protocol (FLDM-00103, Fluidigm, San Francisco, CA, USA). The RT was prepared by mixing 5 μL of purified RNA and 1.25 μL RT Master Mix (100-6297, Fluidigm, San Francisco, CA, USA), followed by 3 steps of incubation using the Bio-Rad T100 thermal cycler (Bio-Rad, Hercules, CA, USA). Preamplification of cDNA was performed by first pooling the N1, N2, and RP assays and diluting into dilution reagent (100-8730, Fluidigm, San Francisco, CA, USA) to a final concentration of 100.5 and 25.5 nM for the primers and probes, respectively, and then, mixing it with 2.5 μL Preamp Master Mix (100-5744, Fluidigm, San Francisco, CA, USA) and 0.635 μL PCR water (100-5941, Fluidigm, San Francisco, CA, USA). The final preamplification reaction (12.5 μL) consisted of preamplification pre-mix and RT reaction in a 1:1 ratio. Upon completion of 20 preamplification cycles, reactions were diluted 1:5 in Dilution Reagent (100-8730, Fluidigm, San Francisco, CA, USA), resulting in a total volume of 62.5 μL.

The qPCR mix was prepared using 1.8 μL of diluted cDNA or SARS-CoV-2 plasmid positive controls (with or without preamplification) and 2 μL of 2X TaqMan Fast Advanced Master Mix (4444557, Thermo Fisher Scientific, Waltham, MA, USA) and 0.2 μL 20X GE Sample Loading Reagent (Fluidigm PN 100-7610). Next, 3 μL qPCR mix from each sample was loaded into the sample inlet in the 192.24 integrated fluid circuit (IFC, Fluidigm, see Figure S1). For each assay, 3 μL of primer/probe mix (13.5X) was mixed with 1 μL of 4X Assay Loading Reagent (Fluidigm, 102-0135), and 3 μL of each assay mixed was loaded in the assay inlet in the 192.24 IFC chip (Fluidigm, 100-626). The chip was then placed in an IFC controller RX machine to pre-load the samples and the assays, and then, loaded onto the BioMark HD instrument (Fluidigm) for RT-qPCR using 35 cycles. In total, 4608 reactions were performed in each 192.24 IFC chip. The raw amplification data were acquired using the Fluidigm data collection software and analyzed using the Fluidigm Real-Time PCR Analysis software 3.0.2. To calculate the Cq (cycle of quantification) of each sample, a global threshold was automatically calculated and applied to all the samples in the chip. The crossing point at which the amplification curve of each sample crossed the threshold line was determined as the Cq. For comparability with other studies, we note that the Cq is numerically equivalent to the Ct value in our system.

2.4. Clinical Samples and SARS-CoV-2 Clinical Diagnostics

A total of 182 de-identified nasopharyngeal (NP) swab samples (91 positive and 91 negatives for SARS-CoV-2) were obtained from an accredited diagnostic lab. The diagnostic lab used the automated NX-48S Viral RNA Kit (NX-48S, Genolution Inc., Seoul, Korea) for RNA extraction, and the U-TOP^{TM} COVID-19 Detection Kit (SS-9930, Seasun Biomaterials Inc., Daejeon, Korea) for SARS-CoV-2 detection, following accredited protocols in the United Arab Emirates that follow guidelines from the US Center for Disease Control and Prevention (CDC). Briefly, three primers were used, two targeting the viral ORF1ab and the N genes, and one acting as an internal control. A sample was classified as positive if at least one of the viral genes was detected with a Ct value ≤38. The reported LoD of the kit was 10 copies/reaction, translating to 1 copy/μL in the RNA sample.

2.5. Three-Step Analysis of SARS-CoV-2 in Clinical Samples

Automated extraction of viral RNA from the clinical samples was performed using the Chemagic 360 automated nucleic acid extraction system (2024-0020, Perkin Elmer, Waltham, MA, USA) and the Chemagic Viral DNA/RNA 300 Kit H96 (CMG-1033S, Perkin Elmer, Waltham, USA) according to the manufacturer's instructions. For RNA extraction, 300 µL clinical samples were used and eluted in 80 µL elution buffers. Subsequent RT, preamplification, and qPCR were performed as described above. Clinical samples were loaded onto two 96-well plates for RT and preamplification, with each plate containing 10-fold serially diluted SARS-CoV-2 plasmid controls (50–5000 copies/µL in the original experiment; 10–10,000 copies/µL in the replication experiments) used for viral load estimation. Samples were considered valid if the RP gene was reliably detected in at least 4 of the 6 replicates. Samples were classified as positive if at least one of the N assays (N1 or N2) was detected in at least one replicate. Each PCR plate also contained two negative controls: empty transport medium, to control for contamination during RNA extraction (NRX control), and TE (Tris–EDTA) dilution buffer, to control for contamination during pre-amplification and qPCR (NQF control).

2.6. Statistical Analysis

Data analysis was performed with R, associated packages, and GraphPad Prism 8. Data were summarized as mean ± standard deviation. Virus copies were quantified based on a 10-fold dilution series of SARS-CoV-2 plasmids to generate standard curves. The standard curves were used to build log-linear models used to predict viral loads based on Cq values.

3. Results

The objective of this study was to evaluate whether a microfluidic nano-scale RT-qPCR system has the potential to enhance the limit of detection (LoD) of SARS-CoV-2 in clinical samples. To do this, we first used synthetic SARS-CoV-2 RNA (Twist RNA) and SARS-CoV-2 plasmids to develop and evaluate our protocols (Figure 1A), and subsequently applied the method to 182 NP samples that were analyzed in a clinical diagnostic lab using standard RT-PCR protocols (Figure 1B). For all our experiments, we used a 192.24 microfluidic chip, which allowed 192 samples to be independently analyzed against 24 different qPCR probes (24 assays), totaling 4608 nano-scale qPCR reactions per experiment (Figure S1A). Following the standards set by the US Center for Disease Control (CDC), we used two probes targeting different regions of the N gene (N1 and N2) for SARS-CoV-2 detection (Figure S1B), and a probe targeting the human Ribonuclease P (RP) gene as a quality control. To increase the robustness of detection for each assay, we performed nine technical replicates for the N1 and N2 assays, and six replicates for the RP assay.

We first used 10-fold dilution series of the synthetic SARS-CoV-2 RNA (Twist RNA) and SARS-CoV-2 plasmid to determine the LoD for the N1 and N2 assays, with and without a preamplification step (Figure 2). With preamplification, the viral N gene was detectable in Twist RNA at 0.5 copies/µL (N1 assay, 9 out of 18 replicates) and 5 copies/µL (N2 assay, 18 out of 18 replicates), whereas without preamplification, we were only able to detect viral material in a few replicates at 5000 copies/µL (N1 assay, 3 out of 18 replicates; N2 assay, 2 out of 18 replicates). No viral material was detected below 5000 copies/µL (Figure 2A). We observed similar results using the SARS-CoV-2 plasmid (Figure 2B), demonstrating that the preamplification step is essential for high detection sensitivity for SARS-CoV-2 in the range of 1 copy/µL. For this reason, we used a preamplification step in all subsequent experiments.

Since the LoD is dependent on the abundance of input material, we next examined the extent to which RNA extraction affects the quantity of input material. We extracted RNA from the Twist RNA dilution series and eluted it in the same volume as the original samples (Figure S2). With or without extraction, we were able to detect 5 copies/µL using both the N1 and N2 assays, but the Cq values for the extracted samples increased by 1–3 cycles at different dilutions. Most viral nucleic acid extraction kits recommend using carrier RNA to enhance recovery of viral RNA in samples where the quantity of material is low; however, carrier RNA might compete nonspecifically with the SARS-CoV-2 RNA in reverse transcription reactions. To explore this possibility, we added carrier RNA directly into the Twist RNA serial dilutions and found that Cq values indeed increased by 0.9–2.7 cycles at most concentrations (Figure S2). This suggests that the presence of carrier RNA, and not RNA extraction itself, can adversely affect the detection of viral material, possibly by interfering with the efficiency of reverse transcription.

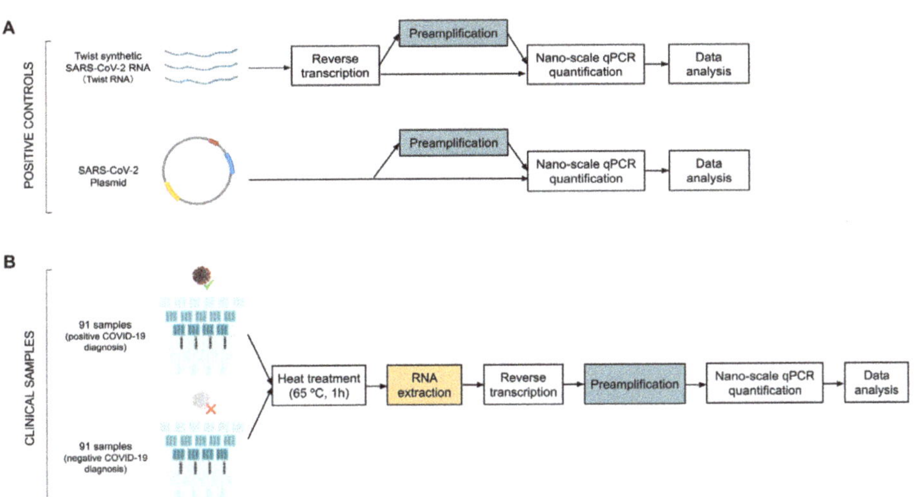

Figure 1. Schematic overview of the experimental design for SARS-CoV-2 RT-qPCR testing using the Fluidigm platform. (**A**) Workflow for method validation in positive controls (SARS-CoV-2 synthetic RNA and SARS-CoV-2 plasmids) with and without a preamplification step. (**B**) Workflow for clinically diagnosed SARS-CoV-2 samples (nasopharyngeal swabs) from an accredited diagnostic lab.

To evaluate the sensitivity of our workflow in comparison with standard SARS-CoV-2 detection methods, we next analyzed 182 NP swab samples previously diagnosed by an accredited diagnostic laboratory as SARS-CoV-2 positive or negative (91 samples each). We heated the samples at 65 °C for 1 h to inactivate viral particles and analyzed them using our SARS-CoV-2 detection workflow (Figure 1B). Any samples with poor qPCR amplification curves and high variation in Cq values among replicates (SEM > 0.5) were flagged as inconsistent, which in our experience is usually due to technical issues such as formation of air bubbles when loading the samples onto the chip. A total of 11 inconsistent samples (4 positive and 7 negative) were removed from subsequent analysis (Figure 3A). The remaining 171 high quality and valid samples were classified as either positive (either N1 or N2 detected) or negative (neither N1 nor N2 detected). Using this classification, we confirmed 86 positive (Pos_Pos, 94.5%) and found 1 negative sample (Pos_Neg, 1.1%) among the 87 samples with a positive clinical diagnosis (Figure 3B). The one Pos_Neg sample had a relatively high Cq value (36.4 for the N-gene and 37.2 for the ORF1 gene) in the clinical diagnostic test; therefore, it is likely that RNA degradation during transport or the heat inactivation process may have brought its viral load below the LoD. Exploring how the two methods compared, we found relative consistency between the ranking of Cq values between the microfluidic RT-qPCR test and the clinical diagnostic test, despite the expected difference in Cq values due to the addition of a pre-amplification step (20 cycles) in the microfluidic method (Figure S3). Among the 84 samples with a negative diagnosis, we detected 17 positives (Neg_Pos, 18.7%): in 9 of these samples, we obtained valid Cq values for all 18 replicates (both N1 and N2), and in the remaining 9, we detected the virus in 9/18 replicates (either all N1 or all N2). All samples with a negative clinical diagnosis showed no detectable signal in the clinical diagnostic qPCR test. These analyses thus show that by performing a large number of replicates, our method can robustly and consistently detect SARS-CoV-2 and identify samples as false-negatives by the clinical diagnostic procedure.

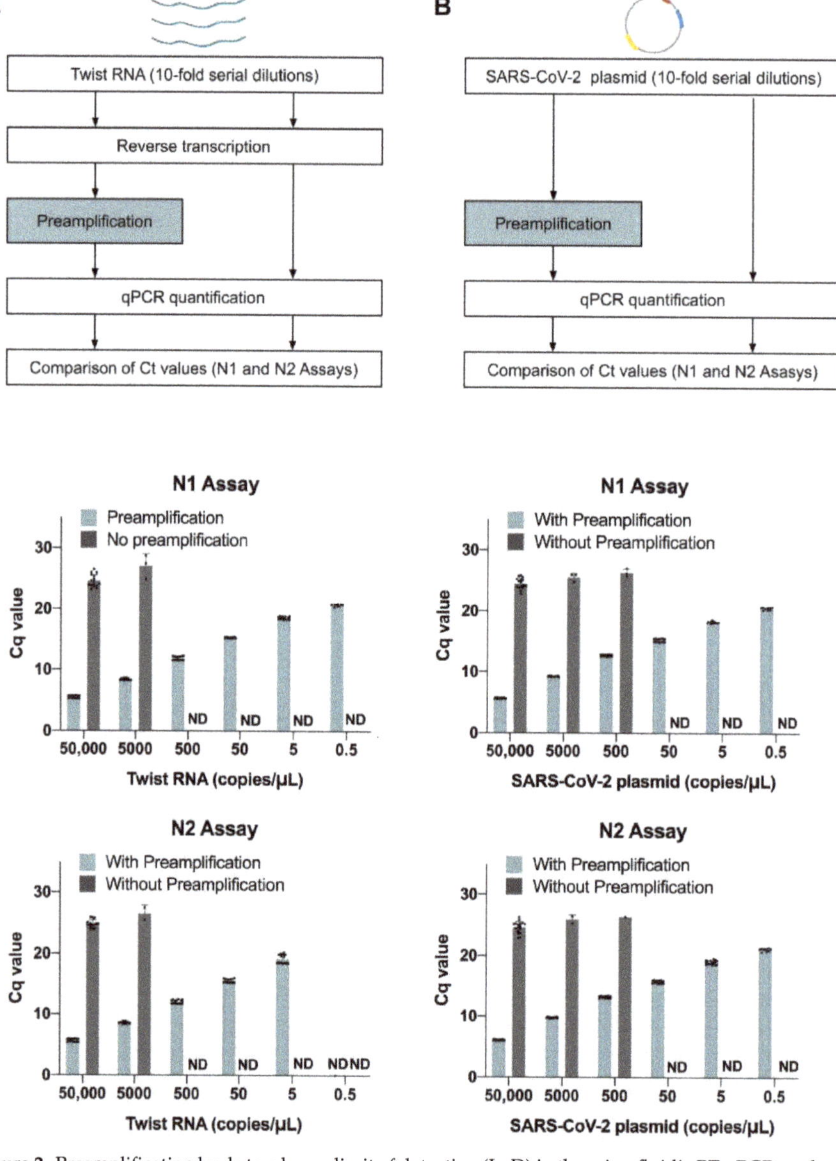

Figure 2. Preamplification leads to a lower limit of detection (LoD) in the microfluidic RT-qPCR method. Workflow for 10-fold serially diluted SARS-CoV-2 positive controls, with and without preamplification: (**A**) synthetic RNAs (Twist RNA) and (**B**) SARS-CoV-2 plasmid. Barplots show the Cq values with and without the preamplification step for both the N1 and N2 assays at each concentration. Data show the mean ± SD of valid Cq values from 18 replicates (2 biological replicates at each concentration, each with 9 technical replicates; each dot represents a technical replicate with a detected Cq value).

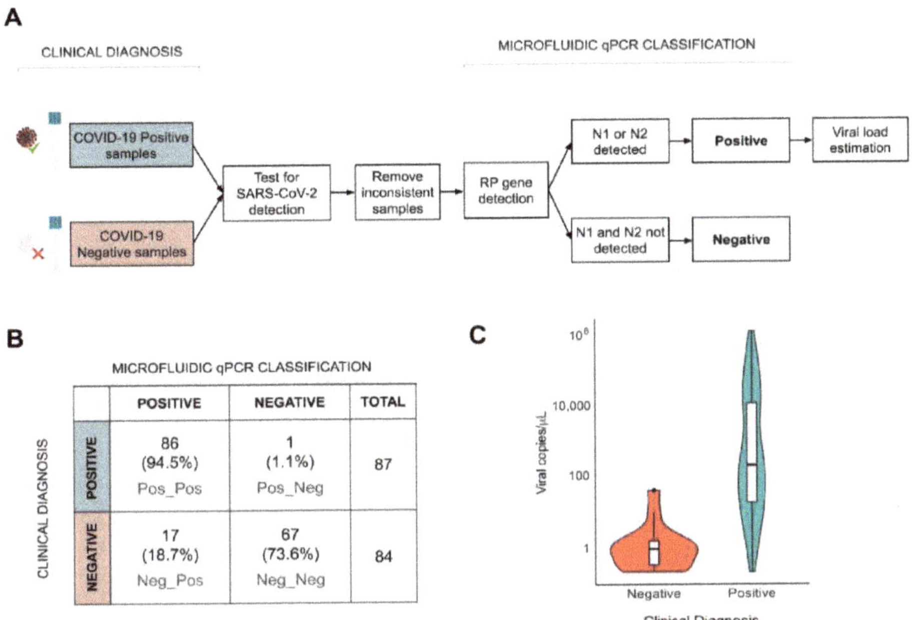

Figure 3. The three-step nano-scale RT-qPCR method shows higher sensitivity in detecting samples with low viral-loads. (**A**) Workflow for the re-analysis of clinical samples (SARS-CoV-2 positive and negative) from an accredited diagnostic lab. (**B**) Comparison between clinical diagnosis and the microfluidic qPCR classification. Category labels indicate clinical diagnosis followed by microfluidic qPCR classification (e.g., Neg_Pos = negative diagnosis, positive microfluidic qPCR result). (**C**) Viral load estimates for different categories of classified samples based on microfluidic qPCR. Samples with a negative clinical diagnosis contained very low viral loads (<100 copies/µL). Viral titers of samples in plate 1 and 2 were computed using the standard curve of plasmid controls from plate 1.

Standard curves are commonly used to estimate viral loads in RT-qPCR reactions, which is why we took that approach in our study. Standard curves based on 100-fold serial dilutions of Twist RNA and SARS-CoV-2 plasmids ranging from 5 to 50,000 copies/µL both showed a nearly-perfect log-linear fit ($R^2 > 0.99$) with little variation among technical replicates (SEM < 0.2) for both N assays (Figure S4). Having established technical precision using these curves, we used the SARS-CoV-2 plasmid standards to quantify viral copies in the clinical samples in each of the two PCR plates. Given that the N1 and N2 assays were highly concordant ($R^2 = 0.876$, Kendall's Tau correlation), and the difference in Cq values was within one cycle in over 90% of the samples (Figure S5A,B), we used an average Cq value for the N gene assays. To maximize consistency in viral load estimates among the samples, we then used the standard curve from plate 1 to quantify the viral loads in each clinical sample (Figure 3C). The Pos_Pos samples showed a wide range of estimated viral loads (0.2–1.17 $\times 10^6$ viral copies/µL) spanning five orders of magnitude. In contrast, the Neg_Pos samples exhibited a much narrower range of viral load estimates (0.2–40.25 viral copies/µL), corresponding to very low amounts of viral material in the NP swab sample (0.05–10.73 viral copies/µL). These results show that the nano-scale qPCR method can detect SARS-CoV-2 across a broad range of viral titers and can confidently detect relatively low viral loads that could otherwise be missed by standard detection methods used in diagnostic labs.

To evaluate the reproducibility of this method in samples with low viral loads, we re-analyzed the 17 Neg_Pos samples after one and after two freeze cycles. After one freeze cycle, 11 samples remained positive, whereas after two freeze cycles, only 9 did (Figure 4A). This suggests that additional freeze cycles may lead to an increased false-negative rate, likely due to degradation of viral RNA resulting in copy numbers below the LoD. This is consistent with prior observations of increased Cq values following one freeze–thaw cycle [27]. Nevertheless, 8 of the 17 samples were reproducibly classified as positive in three independent experiments (Figure 4B). When comparing viral load estimates among the 17 Neg_Pos samples, we found that the 11 samples that retested positive after one freeze cycle had viral loads between 0.21 and 38.89 copies/μL, whereas the 5 samples that retested negative all had viral loads less than 1 copy/μL based on the original experiment (Figure 4C). This suggests that samples with extremely low viral titers close to the LoD could fail to be consistently detected, likely due to factors such as sample degradation or stochastic variation in the number of viral RNA molecules present in the small reaction volumes used for RT.

A

	Original experiment	After one freeze cycle	After two freeze cycles
Positive	17	11	9
Negative	0	5	5
Inconsistent	0	1	3

B

After one freeze cycle \ After two freeze cycles	Positive	Negative	Inconsistent
Positive	8	2	1
Negative	1	2	2
Inconsistent	0	1	0

C

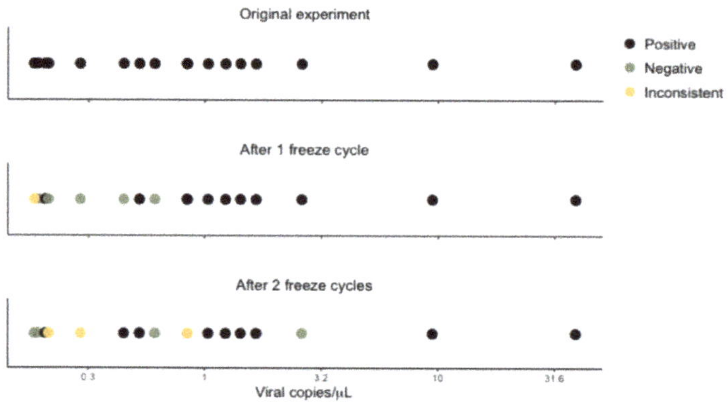

Figure 4. Reproducibility of results among the 17 Neg_Pos samples. (**A**) Summary of the microfluidic qPCR results across the original experiment and two replication experiments (after one freeze cycle and after two freeze cycles). (**B**) Cross-tabulation of microfluidic qPCR results in the two replication experiments. (**C**) Viral load estimates for the 17 Neg_Pos samples based on the original experiment. Color coding is based on the microfluidic qPCR results of the 17 samples in the two replication experiments, as either positive (black), negative (green), or inconsistent (yellow).

Lastly, we examined the relationship between the distributions of mean Cq values for the host RP gene assay and the N gene assay. If negative samples tended to show much higher Cq values for the RP assay than the positive samples, this would indicate that the negative result was likely due to inadequate sampling of human tissues in the swab. Instead, we observed no significant differences in the Cq values of RP assay between positive and negative samples (t-test, p-value = 0.08) (Figure 5A) and found minimal correlation between the mean Cq values of the RP and the N gene assays, with only 2.8% of the variation in the N gene assays attributable to detection of the RP gene ($R^2 = 0.028$, Kendall's Tau correlation) (Figure 5B). We thus conclude that the high Cq values for the N assays in the Neg_Pos samples were not due to inadequate sampling, but instead, accurately reflected low viral loads in these samples.

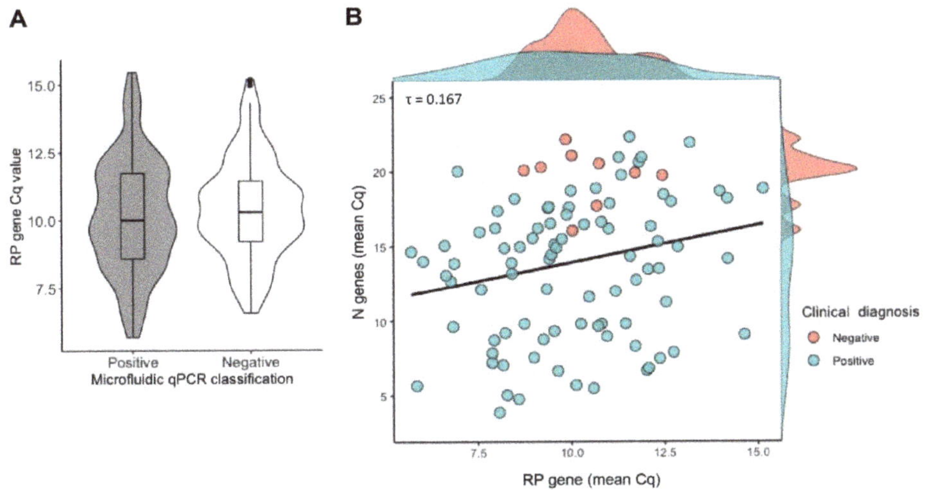

Figure 5. Low estimated viral titers are not due to poor quality samples. (**A**) Violin plots of RP gene Cq values across samples with positive and negative microfluidic qPCR results. There is no significant difference between the two groups (Welch's *t*-test, *p*-value = 0.08). (**B**) Scatter and density plots of N gene vs. RP gene Cq values from microfluidic qPCR for samples with negative or positive clinical diagnostic results. Cq values for viral N and host RP genes are not strongly correlated overall ($\tau = 0.167$, *p*-value = 0.017).

4. Discussion

In this study, we implement and validate a three-step approach for SARS-CoV-2 detection utilizing RT of SARS-CoV-2 viral RNA, cDNA preamplification, and nano-scale qPCR. Using serial dilutions of positive controls, we demonstrate a 1000-fold improvement in detection sensitivity of the microfluidic qPCR system when adding the preamplification step, consistent with a previous study in which a target preamplification step was found to enhance the LoD by 100-fold in detecting SARS-CoV [23]. We also show that nano-scale qPCR can be used to quantify viral copies in clinical samples with high confidence. Our data suggest that the LoD of this method (with preamplification) is less than 1 copy/μL: we obtained a LoD of 0.5 copies/μL for Twist RNA and SARS-CoV-2 plasmid, and detected the virus down to 0.2 copies/μL in experiments using clinical NP swab samples. Based on this analysis, our method seems to be more sensitive than standard RT-PCRs, which have reported a LoD ranging between 5.6 and 100 copies/μL [28–30], and that its performance is comparable to other highly sensitive methods such as the CDC 2019-nCoV RT-PCR Diagnostic Panel with QIAGEN QIAmp DSP Viral RNA Mini Kit and the QIAGEN EZ1 DSP (1 copy/μL) [3], ddPCR (0.1 copies/μL) [8], and RT-LAMP (1 copy/μL) [31].

Crucially, we demonstrate the power of this method to reduce the false-negative rate of SARS-CoV-2 clinical diagnostic tests: we detected SARS-CoV-2 in 17 samples diagnosed as negative by an accredited diagnostic lab with high confidence, based on a large number of technical replicates (9 for N1, 9 for N2, 6 for RP) and a conservative threshold for Cq value consistency among sample replicates (SEM < 0.5). This is especially important considering that the viral loads of these Neg_Pos samples (0.2–40.25 viral copies/μL) were close to the LoD of standard SARS-CoV-2 tests and are thus, more likely to return false-negative results using standard RT-PCR methods. This three-step microfluidics RT-qPCR method, which includes a preamplification step, nano-scale reactions, and a large number of replicates, can reliably detect samples with low viral load and reduce the false-negative rate compared to standard RT-PCR assays.

Beyond its ultra-sensitive detection of SARS-CoV-2, this microfluidic platform is a cost-effective strategy with several advantages: a nanoliter volume per reaction (lower reagent consumption per assay), a parallelized assay system (increased throughput), amenability to automation (increased precision), capacity to run a large number of replicates per sample (increased confidence in test results), and the capacity to simultaneously test for multiple pathogens (broader diagnostic utility). Based on our experience, and considering the fact that reagents cost more in the UAE than in Europe and the US, the cost of SARS-CoV-2 testing per sample using the 192.24 microfluidic chip and the design used in this study (9 replicates for N1 and N2, 6 replicates for RP) is approximately USD 17, including reagents and the chip, but excluding labor costs. This is approximately four times cheaper than the current cost of standard SARS-CoV-2 testing in the UAE. Lowering the number to three technical replicates for each of the N1, N2, and RP assays would allow for four additional assays, which can be utilized to diagnose other viral or bacterial infections using three technical replicates (or six additional assays with two technical replicates) at no extra cost, bringing down the cost per sample per assay even further. Since each sample is loaded only once and then analyzed against 24 assays simultaneously, no extra time is required for performing multiple assays of replicates. Because of this, the method has great potential for economies of scale, including assay multiplexing (to detect additional pathogens) and sample pooling (to increase throughput), which would further reduce per test costs. These advantages warrant serious consideration of this three-step nano-scale assay system, especially for active screening programs, which aim at the early detection of SARS-CoV-2 in asymptomatic, pre-asymptomatic, or mildly symptomatic individuals who likely carry low viral loads.

Supplementary Materials: The following are available online at http://www.mdpi.com/2227-9717/8/11/1425/s1, Figure S1: Fluidigm chip layout and N gene probes; Figure S2: Detection limit with RNA extraction step and the effect of carrier RNA; Figure S3: Consistency in Cq value ranking between the microfluidic RT-qPCR and the clinical diagnostic method; Figure S4: Linear regression of Cq values vs. copy number of serially diluted positive controls shows highly reproducible detection of SARS-CoV-2 across five orders of magnitude; Figure S5: Comparison of detection performance between N1 and N2 assays.

Author Contributions: Conceptualization, K.C.G., Y.I., and F.P.; methodology, X.X., Z.A., M.M.D., M.A., M.K., Y.M., F.A.J., and K.P.; software K.C.G., N.R., and C.A.J.; resources, L.E.M., Z.V., and M.Z.; formal analysis, T.G. and X.X.; writing—original draft preparation, X.X., T.G., K.C.G., Y.I., Z.A., and M.M.D., writing—review and editing, X.X., T.G., K.C.G., Y.I., Z.A., and M.M.D., funding acquisition, F.P., K.C.G., Y.I., and R.A. All authors have read and agreed to the published version of the manuscript.

Funding: This work was funded by support from NYU Abu Dhabi to the NYUAD Center for Genomics and Systems Biology (NYUAD Research Institute grant #ADHPG-CGSB1 to K.C.G.), NYUAD research grant AD105 (to Y.I.), the NYUAD Kawader Research Assistantship Program (to F.A.J.), and the NYUAD Core Technology Platforms.

Acknowledgments: This report is part of NYUAD's COVID Response Project. We thank members of NYU Abu Dhabi's (NYUAD) COVID-19 steering committee for spearheading and facilitating this project, including Ayaz Virji and Sehamuddin Galadari. We thank members of Proficiency Healthcare Diagnostics laboratory for sample collection and processing. We thank Michael Davis (Director of Laboratory Operations, NYUAD), Reza Rowshan (Director of Core Technology Platforms, NYUAD), Nizar Drou (Lead Developer, NYUAD Bioinformatics Core), Nada Messaikeh (Vice Provost for Research Administration, NYUAD), and Enas Qudeimat (Director of Operations, NYUAD-CGSB) for helping to make this work possible. We thank the Institutional Review Boards of the Abu Dhabi Department of Health and NYUAD and the NYUAD Institutional Biosafety Committee for expediting the review of this project and NYUAD Environmental Health and Safety for their support. This research was partially carried out using the Core Technology Platforms resources and the High-Throughput Screening (HTS) platform of CGSB at New York University Abu Dhabi.

Conflicts of Interest: The authors declare no conflict of interest. The funders had no role in the design of the study; in the collection, analyses, or interpretation of data; in the writing of the manuscript, or in the decision to publish the results.

References

1. World Health Organization (WHO). Laboratory Testing for 2019 Novel Coronavirus (2019-nCoV) in Suspected Human Cases, Interim Guidance, 2 March 2020. Available online: https://apps.who.int/iris/bitstream/handle/10665/331329/WHO-COVID-19-laboratory-2020.4-eng.pdf?sequence=1&isAllowed=y (accessed on 28 July 2020).
2. Corman, V.M.; Landt, O.; Kaiser, M.; Molenkamp, R.; Meijer, A.; Chu, D.K.; Bleicker, T.; Brünink, S.; Schneider, J.; Schmidt, M.L.; et al. Detection of 2019 novel coronavirus (2019-nCoV) by real-time RT-PCR. *Eurosurveillance* **2020**, *25*, 2000045. [CrossRef] [PubMed]
3. Centers for Disease Control and Prevention. Real-Time RT–PCR Panel for Detection 2019-nCoV (US Centers for Disease Control and Prevention, 2020). Available online: https://www.fda.gov/media/134922/download (accessed on 28 July 2020).
4. Dramé, M.; Teguo, M.T.; Proye, E.; Hequet, F.; Hentzien, M.; Kanagaratnam, L.; Godaert, L. Should RT-PCR be considered a gold standard in the diagnosis of COVID-19? *J. Med. Virol.* **2020**. [CrossRef]
5. Zhang, J.-F.; Yan, K.; Ye, H.-H.; Lin, J.; Zheng, J.-J.; Cai, T. SARS-CoV-2 turned positive in a discharged patient with COVID-19 arouses concern regarding the present standards for discharge. *Int. J. Infect. Dis.* **2020**, *97*, 212–214. [CrossRef] [PubMed]
6. Xiao, A.T.; Tong, Y.X.; Zhang, S. False negative of RT-PCR and prolonged nucleic acid conversion in COVID-19: Rather than recurrence. *J. Med. Virol.* **2020**, *92*, 1755–1756. [CrossRef]
7. Wikramaratna, P.; Paton, R.S.; Ghafari, M.; Lourenco, J. Estimating false-negative detection rate of SARS-CoV-2 by RT-PCR. *medRxiv* **2020**. [CrossRef]
8. Lu, R.; Wang, J.; Li, M.; Wang, Y.; Dong, J.; Cai, W. SARS-CoV-2 detection using digital PCR for COVID-19 diagnosis, treatment monitoring and criteria for discharge. *medRxiv* **2020**. [CrossRef]
9. Yu, F.; Yan, L.; Wang, N.; Yang, S.; Wang, L.; Tang, Y.; Gao, G.; Wang, S.; Ma, C.; Xie, R.; et al. Quantitative Detection and Viral Load Analysis of SARS-CoV-2 in Infected Patients. *Clin. Infect. Dis.* **2020**, *71*, 793–798. [CrossRef]
10. Guo, Y.; Wang, K.; Zhang, Y.; Zhang, W.; Wang, L.; Liao, P. Comparison and analysis of the detection performance of six new coronavirus nucleic acid detection reagents. *Chongqing Med.* **2020**, *14*, 1671–8348.
11. Kapitula, D.S.; Jiang, Z.; Jiang, J.; Zhu, J.; Chen, X.; Lin, C.Q. Performance & Quality Evaluation of Marketed COVID-19 RNA Detection Kits. *medRxiv* **2020**. [CrossRef]
12. Eisen, A.K.A.; Demoliner, M.; Gularte, J.S.; Hansen, A.W.; Schallenberger, K.; Mallmann, L.; Hermann, B.S.; Heldt, F.H.; De Almeida, P.R.; Fleck, J.D.; et al. Comparison Of Different Kits For SARS-CoV-2 RNA Extraction Marketed In Brazil. *bioRxiv* **2020**. [CrossRef]
13. Furukawa, N.W.; Brooks, J.T.; Sobel, J. Evidence Supporting Transmission of Severe Acute Respiratory Syndrome Coronavirus 2 While Presymptomatic or Asymptomatic. *Emerg. Infect. Dis.* **2020**, *26*. [CrossRef]
14. Gandhi, M.; Yokoe, D.S.; Havlir, D.V. Asymptomatic Transmission, the Achilles' Heel of Current Strategies to Control Covid-19. *N. Eng. J. Med.* **2020**, *382*, 2158–2160. [CrossRef] [PubMed]
15. Bai, Y.; Yao, L.; Wei, T.; Tian, F.; Jin, D.-Y.; Chen, L.; Wang, M. Presumed Asymptomatic Carrier Transmission of COVID-19. *JAMA* **2020**, *323*, 1406–1407. [CrossRef] [PubMed]
16. Rothe, C.; Schunk, M.; Sothmann, P.; Bretzel, G.; Froeschl, G.; Wallrauch, C.; Zimmer, T.; Thiel, V.; Janke, C.; Guggemos, W.; et al. Transmission of 2019-nCoV Infection from an Asymptomatic Contact in Germany. *N. Eng. J. Med.* **2020**, *382*, 970–971. [CrossRef]
17. Wang, C.; Liu, L.; Hao, X.; Guo, H.; Wang, Q.; Huang, J.; He, N.; Yu, H.; Lin, X.; Pan, A.; et al. Evolving epidemiology and impact of non-pharmaceutical interventions on the outbreak of Coronavirus disease 2019 in Wuhan, China. *medRxiv* **2020**. [CrossRef]
18. Ing, A.J.; Cocks, C.; Green, J.P. COVID-19: In the footsteps of Ernest Shackleton. *Thorax* **2020**, *75*, 693–694. [CrossRef] [PubMed]
19. Oran, D.P.; Topol, E.J. Prevalence of Asymptomatic SARS-CoV-2 Infection. *Ann. Intern. Med.* **2020**, *173*, 362–367. [CrossRef] [PubMed]
20. Li, R.; Pei, S.; Chen, B.; Song, Y.; Zhang, T.; Yang, W.; Shaman, J. Substantial undocumented infection facilitates the rapid dissemination of novel coronavirus (SARS-CoV-2). *Science* **2020**, *368*, 489–493. [CrossRef] [PubMed]

21. Zhou, R.; Li, F.; Chen, F.; Liu, H.; Zheng, J.; Lei, C.; Wu, X. Viral dynamics in asymptomatic patients with COVID-19. *Int. J. Infect. Dis.* **2020**, *96*, 288–290. [CrossRef]
22. Kucirka, L.M.; Lauer, S.A.; Laeyendecker, O.; Boon, D.; Lessler, J. Variation in False-Negative Rate of Reverse Transcriptase Polymerase Chain Reaction–Based SARS-CoV-2 Tests by Time Since Exposure. *Ann. Intern. Med.* **2020**, *173*, 262–267. [CrossRef]
23. Lau, L.T.; Fung, Y.-W.W.; Wong, F.P.-F.; Lin, S.S.-W.; Wang, C.R.; Li, H.L.; Dillon, N.; A Collins, R.; Tam, J.S.-L.; Chan, P.; et al. A real-time PCR for SARS-coronavirus incorporating target gene pre-amplification. *Biochem. Biophys. Res. Commun.* **2003**, *312*, 1290–1296. [CrossRef] [PubMed]
24. Parker, J.K.; Chang, T.-Y.; Meschke, J.S. Amplification of viral RNA from drinking water using TransPlex™ whole-transcriptome amplification. *J. Appl. Microbiol.* **2011**, *111*, 216–223. [CrossRef]
25. Jackson, J.B.; Choi, D.S.; Luketich, J.D.; Pennathur, A.; Ståhlberg, A.; Godfrey, T.E. Multiplex Preamplification of Serum DNA to Facilitate Reliable Detection of Extremely Rare Cancer Mutations in Circulating DNA by Digital PCR. *J. Mol. Diagn.* **2016**, *18*, 235–243. [CrossRef] [PubMed]
26. Del Gaudio, S.; Cirillo, A.; Di Bernardo, G.; Galderisi, U.; Thanassoulas, T.; Pitsios, T.; Cipollaro, M. Preamplification Procedure for the Analysis of Ancient DNA Samples. *Sci. World J.* **2013**, *2013*, 1–8. [CrossRef]
27. Li, L.; Li, X.; Guo, Z.; Wang, Z.; Zhang, K.; Li, C.; Wang, C.; Zhang, S. Influence of Storage Conditions on SARS-CoV-2 Nucleic Acid Detection in Throat Swabs. *J. Infect. Dis.* **2020**, *222*, 203–205. [CrossRef] [PubMed]
28. Han, M.S.; Seong, M.-W.; Kim, N.; Shin, S.; Cho, S.I.; Park, H.; Kim, T.S.; Park, S.S.; Choi, E.H. Viral RNA Load in Mildly Symptomatic and Asymptomatic Children with COVID-19, Seoul. *Emerg. Infect. Dis.* **2020**, *26*, 2497–2499. [CrossRef]
29. Wyllie, A.L.; Fournier, J.; Casanovas-Massana, A.; Campbell, M.; Tokuyama, M.; Vijayakumar, P.; Geng, B.; Muenker, M.C.; Moore, A.J.; Vogels, C.B.F.; et al. Saliva is more sensitive for SARS-CoV-2 detection in COVID-19 patients than nasopharyngeal swabs. *medRxiv* **2020**. [CrossRef]
30. Vogels, C.B.F.; Brito, A.F.; Wyllie, A.L.; Fauver, J.R.; Ott, I.M.; Kalinich, C.C.; Petrone, M.E.; Casanovas-Massana, A.; Muenker, M.C.; Moore, A.J.; et al. Analytical sensitivity and efficiency comparisons of SARS-COV-2 qRT-PCR assays. *medRxiv* **2020**. [CrossRef]
31. Yang, W.; Dang, X.; Wang, Q.; Xu, M.; Zhao, Q.; Zhou, Y.; Zhao, H.; Wang, L.; Xu, Y.; Wang, J.; et al. Rapid detection of SARS-CoV-2 using reverse transcription RT-LAMP method. *medRxiv* **2020**. [CrossRef]

Publisher's Note: MDPI stays neutral with regard to jurisdictional claims in published maps and institutional affiliations.

MDPI
St. Alban-Anlage 66
4052 Basel
Switzerland
Tel. +41 61 683 77 34
Fax +41 61 302 89 18
www.mdpi.com

Processes Editorial Office
E-mail: processes@mdpi.com
www.mdpi.com/journal/processes

www.ingramcontent.com/pod-product-compliance
Lightning Source LLC
LaVergne TN
LVHW070631170726
843515LV00004B/225

* 9 7 8 3 0 3 6 5 1 3 6 6 9 *